U.S. Housing Policy, Politics, and Economics

The stirrings of reform or more of the same? *U.S. Housing Policy, Politics, and Economics* shares a stark and urgent message. With a new president in the White House and the economy emerging from its peak pandemic lows, the time is right for transformative federal housing legislation – but only if Congress can transcend partisan divides. Drawing on nearly a century of legislative and policy data, this briefing for scholars and professionals quantifies the effects of Democratic or Republican control of the executive and legislative branches on housing prices and policies nation-wide. It exposes the lasting consequences of Congress' more than a decade of failure to pass meaningful housing laws and makes clear just how narrow the current window for action is. Equal parts analysis and call to arms, *U.S. Housing Policy, Politics, and Economics* is essential reading for everyone who cares about affordable, accessible housing.

Lawrence A. Souza, DBA, participated in underwriting more than $900 million in real estate development and acquisitions deals in his commercial real estate career. That experience informs his current role as a real estate economist and adjunct professor at Saint Mary's College of California.

Hannah Macsata earned her BS in Business Administration with a concentration in Marketing from Saint Mary's College of California.

Dustin Hartuv earned his BA in Economics at Georgetown University and is currently a JD candidate at Cornell Law School.

Joshua Martinez is entering his junior year at Saint Mary's College of California. He is majoring in Kinesiology with a concentration in Sports Management.

Alicia Bilbrey-Becker earned her BS in Business Administration with a concentration in Marketing from Saint Mary's College of California.

U.S. Housing Policy, Politics, and Economics

Bias and Outcomes

Lawrence A. Souza,
Hannah Macsata, Dustin Hartuv,
Joshua Martinez, and
Alicia Bilbrey-Becker

Routledge
Taylor & Francis Group

NEW YORK AND LONDON

First published 2022
by Routledge
605 Third Avenue, New York, NY 10158

and by Routledge
2 Park Square, Milton Park, Abingdon, Oxon, OX14 4RN

Routledge is an imprint of the Taylor & Francis Group, an informa business

© 2022 Lawrence A. Souza, Hannah Macsata, Dustin Hartuv, Joshua Martinez, and Alicia Bilbrey-Becker

Library of Congress Cataloging-in-Publication Data
A catalog record for this title has been requested

ISBN: 978-1-032-11483-5 (hbk)
ISBN: 978-1-032-12176-5 (pbk)
ISBN: 978-1-003-22343-6 (ebk)

DOI: 10.1201/9781003223436

Typeset in Times New Roman
by codeMantra

Contents

Acknowledgments

I would like to thank Dustin Hartuv of Georgetown University, Joshua Martinez of Saint Mary's College of California, Hannah Macsata, and Alicia Bilbrey-Becker in helping me research and write this paper; I would not have done it without them. I would also like to acknowledge Dr. John Quigley, Dr. Ken Rosen, and Dr. Bob Edelstein, U.C. Berkeley, they are role models for me, and brilliant housing policy economists that I have tremendous admiration and respect for.

Foreword

It is a daunting task to write a short, concise, accessible analysis to such a complex topic as "housing policy in the United States." This challenge becomes even more formidable in the case of a very short manuscript. It is, therefore, not surprising that the most authors of existing short manuscripts to the subject have opted to discuss only one aspect or perhaps two aspects of housing policy, usually focusing on topics such as housing economics, housing policy history, housing financial institutions, credit markets, and housing mortgage-backed securitization. While these types of studies are helpful for explaining the intricacies of housing policy and housing market outcomes or the financial systems and institutions supporting housing, such narrow accounts often leave the general reader with a shallow understanding of intricacies and interstices of how housing policy has affected housing markets.

The present manuscript makes the case that housing policy is best thought of as a multidimensional set of social, political, financial, and economic processes that resists being confined to any single thematic framework. Indeed, the transformative powers of housing policy reach deeply into the economic, political, cultural, technological, and ecological dimensions of contemporary social life. To be sure, the discussion of economic matters must be a significant part of any comprehensive account for housing policy, but the latter should not be conflated with the former.

U.S. housing policy contains important discursive aspects in the form of ideologically charged narratives that put before the

x *Foreword*

public a particular agenda of topics for discussion, questions to ask, and claims to make. The existence of these narratives shows that housing policy is not merely an objective process, but also a plethora of stories that define, describe, and analyze that very process. The social forces behind these competing accounts of housing policy seek to direct political actions that potentially affect millions of Americans.

This book has been created with a keen awareness that the study of housing policy crosses and intertwines the disciplines of economics, financial institutions, social equity, and politics. The lack of a firm pure disciplinary home for housing policy also contains great opportunity. Housing policy analysis in this context is an emerging field that cuts across traditional disciplinary boundaries. This strong emphasis on multiple disciplines requires students of housing policy to familiarize themselves with literatures on subjects that have often been studied in isolation from each other. A significant challenge confronting housing policy researchers is connecting and synthesizing the various strands of knowledge in a way that places proper weight and emphasis on the various discipline impacts on housing policy. In brief, housing policy, as analyzed in this manuscript, requires an interdisciplinary approach that is broad enough to behold the big picture.

While the main purpose of the book is to provide its audience with a set of descriptive, qualitative, and quantitative explanatory accounts of various dimensions of housing policy, the careful reader will detect throughout the chapters a critical undertone. This book illustrates how political power can be utilized to create stabilizing supply-side policies and augmenting demand-side housing policies that can be utilized to address the growing U.S. housing crises as well as social issues pertaining to housing access inequities. In this way, this book provides an interesting and important way to address housing policy in the future.

Dr. Robert Edelstein
University of California Berkeley, Haas Business School
Professor Emeritus, Maurice Mann Chair in Real Estate,
Co-Chair, Fisher Center for Real Estate & Urban Economics

Edelstein has served as a consultant to the Philadelphia
Finance Department, U.S. Department of Housing and Urban
Development, the Housing Development Corporation of
New York City, the Federal Reserve Bank of Philadelphia,
the Federal Home Loan Bank of San Francisco, and the U.S.
Department of Energy. He served as president of the American
Real Estate and Urban Economics Association in 1996.

1 Introduction

According to Home Budget: Cost-of-Living Reality Check (2019), housing is an essential factor in determining the quality of lives, stability of communities, and health of national economies. Its importance to society is underscored by the fact – in the United States, for example – that housing accounts for roughly 25–30% of personal consumption expenditures and the same proportion of gross private domestic investment. The status of the housing sector is a leading indicator of economic activity, especially in the United States where the health of the housing industry is extremely sensitive to monetary and fiscal conditions and policies.

In the United States and other industrialized countries where housing quality is high, affordability has become a major issue. In developing countries, longstanding problems of low quality and high relative cost have been exacerbated by high rates of population growth and country-to-city migration, and by urban infrastructure ill-equipped to accommodate residential growth.

Direct government assistance for housing in both industrialized and developing countries has been more extensive than in the United States. With the adoption of the Housing Act of 1949, however, the United States formally pledged itself to the goal of providing "a decent home and a suitable living environment for every American family." Nevertheless, the definition of what is "decent" has varied according to economic conditions, political climate, and prevailing tastes. Furthermore, in the United States the responsibility for producing housing and delivering housing services remains almost exclusively in the private sector.

DOI: 10.1201/9781003223436-1

Due to the importance of housing in determining the way people live in different communities of the world, it is important to understand factors that both stabilize and destabilize housing industries and the housing market as a whole. The best way to analyze housing in the United States is through legislation produced by Congress. Congress, as well as presidents through executive orders, have the ability to influence markets through either supply-side or demand-side legislation.

As the names suggest, *supply-side legislation involves policies that promote the supply of housing, while demand-side legislation involves policies that promote demand for housing.* Overall, in regard to the history of U.S. housing policy, *the research proves there has been supply-side legislation bias.* Demand-side policies do not occur often, generally only during recessions. We hypothesize that supply-side legislative bias has led to stable housing markets since the 1950s, allowing most Americans to become homeowners; however, this trend has reversed.

There has been clear supply-side policy bias, as over the 88 years of housing policy in the United States, housing policy is generally created by Democrats, whether it be a Democratic President, Senate, and/or House. Nevertheless, there have been larger policy gaps in overall housing policy recently, and we believe this is the main reason for the current housing crisis. Finally, of the policies that have been enacted in the past, allowing for lower real home prices and higher affordability, 28 of the 35 policies have been supply side, over the 88 years of U.S. housing policy.

Furthermore, housing-related policies have become fewer, with larger policy gap periods, and if this continues in the future, we anticipate a continuation of the housing crisis, reflected in rising prices and historically low affordability. *The majority of all housing policies, both supply side and demand side, have been enacted under a Democratic Party President, Senate, and/or House of Representatives. Democrats are two times more likely to pass housing policies, specifically supply-side policies.*

In the last decade, there has been a shift toward more demand-side legislation, or no legislation (policy gaps). Consequently, fewer supply-side policies have led to rising costs and falling affordability for housing consumers. If this continues, we anticipate a continuation of the housing crisis, reflected in rising prices and historically low affordability. While most demand-side policies occur during recessions, the party in power is more of a determinant of policies being enacted rather than whether there is a recession.

Based on this research, with a Republican-controlled executive branch, congressional Republican Party control of the senate, and democratic majority control of the house (2016 to 2020), there was no real housing policy passed during the two-and-four year congressional and presidential election cycles. With a Democrat elected as president, and a democratically controlled house and senate during the (2021 to 2024) congressional and presidential election cycles, there is a high probability of significant supply-side housing policy enacted. For example, Biden-Harris housing policy initiatives include Section 8 housing choice vouchers, affordable housing construction, mitigation of discriminatory housing policies, reduction in exclusionary zoning policies, higher density housing, etc.

Overall, Democrats have enacted more supply-side housing policy and show a bias when it comes to enacting supply-side housing policy. When comparing party Presidents with their opposing party-controlled House and Senate, a Republican President and Democratic Congress have enacted four supply-side housing policies while only blocking six (67% success ratio). In contrast, under a Democratic Party President, and Republican Congress, only one supply-side housing policy has been enacted while four have been blocked (25% success ratio). Historically, Democrats show a bias toward supply-side housing policy, and have had greater success at enacting supply-side policy under opposing party Presidents.

However, in the *last decade, there has been a shift toward more demand-side legislation, or no legislation (policy gaps) at all*, even though the traditional institutional supply-side policy bias still remains. With fewer supply-side policies, the traditional supply shocks to the housing market have declined, leading to more benefits for housing developers and brokers, but only detriments (rising costs and falling inventory and affordability) for housing consumers. *The scaling back of supply-side legislation must be slowed, or reversed, to ensure the maximum benefits (affordability/ access) for U.S. households.*

The first section of this research discusses aspects of major housing policy in the United States. The second section of this research explores factors driving supply and demand for housing in the United States. Standard Ordinary Least Squares (OLS) regression techniques are used to test the significance of the variables presented. A forecast model for both supply and demand is created, supply of and demand for housing is calculated, and a projected supply-demand gap analysis is conducted. The gap measures the difference between housing production and demand, and long-run implications on home price appreciation and affordability.

The last section discusses the overall housing gap and future housing market conditions. Depending on the degree of gap, supply-side or demand-side policy recommendations are made to correct for the imbalance, and try to bring the market back into macroeconomic equilibrium over the long run.

Lastly, before investors can build and manage institutional grade single-family or apartment portfolios, there needs to be a basic understanding of policy and economic fundamentals that drive demand and supply of housing. Once aggregate factors that drive housing demand are determined, single-family and apartment market supply and demand fundamentals can be analyzed, and factors determined for metro and submarket performance can be used in the selection process, portfolio optimization, and target allocations. These allocations can also be determined, and hedging strategies established, using indices for benchmarking and property derivative positions.

Citations

Worley, J. (2019, October 18). *Home budget: Cost-of-living reality check*. Quicken. https://www.quicken.com/home-budget-cost-living-reality-check

2 Statement of Purposes

Chronology-Evolution Housing Policy-Politics-Economics

The purpose of this section is to discuss significant urban housing policy in the United States, to identify significant macro-level housing supply and demand variables for forecasting, determine the degree of imbalance between supply and demand for housing, the implications of this imbalance, and recommend policies to counteract these imbalances.

Statement of Specific Research Objectives

To find out and analyze:

- The chronology and evolution of significant housing policy and party politics in the United States.
- The drivers of housing demand for forecasting purposes.
- The drivers of housing supply for forecasting purposes.
- The gap between supply and demand for housing in the United States.

DOI: 10.1201/9781003223436-2

3 Literature Review

Eighty-Eight Years of U.S. Housing Policy

History of U.S. Housing Policy

The following section describes significant U.S. housing policy over the past 88 years.

Housing Policy 1800s

According to Fish (1979), in the 19th century, industrialization, improved transportation, and large-scale immigration led to the rapid expansion of urban areas. Between 1840 and 1850 alone, the urban population of the United States increased by almost 10% from 18.5% to 26.9%. In major industrial cities, much of this growth was accommodated in tenement flats and other multi-story structures.

As mentioned in the Social Welfare History Project (2018), despite widespread public recognition of worsening urban housing problems and frequent calls for reform, only after the Civil War were government efforts undertaken to improve housing conditions. *In 1867, the New York state legislature enacted the first tenement-housing legislation, which regulated the construction of railroad flats by establishing minimum construction standards.* The continued influx of immigrants, however, resulted in the proliferation of overcrowded tenements and deplorable health conditions.

Attempts to improve housing were spurred by the writings of such reformers as Jacob Riis (*How the Other Half Lives)* and Lawrence Veiller in the 1890s, as well as the first federal report on

DOI: 10.1201/9781003223436-3

housing conditions, issued in 1894. Moreover, writers of novels like Charles Dickens (*Great Expectations, Bleak House*) and Upton Sinclair (*The Jungle*) discussed themes of crowded cities and tenements in their stories.

Nevertheless, it was not until 1901 that a law permitting enforcement of housing standards was enacted. According to the New York Public Library (2018), the landmark New York City "New Law" required building permits and inspections, prescribed penalties for noncompliance, and created a permanent city housing department. Subsequently, the New Law was copied in other U.S. cities and provided an impetus for housing legislation at the state level in the early 1900s.

During the depression home building was at a standstill and the rate of foreclosures was intolerably high; emergency programs (Reconstruction Finance Program, Works Progress Administration, etc.) were enacted by Hoover and Roosevelt administrations to provide jobs in the construction industry, improve housing conditions, and extend financial assistance to people threatened with the loss of their homes as cited by Markels (2008).

Beginning of Extensive U.S. Housing Policy (The Great Depression)

The first major step toward a national housing policy was the creation in 1932 of the *Federal Home Loan Bank System (FHL-Bank System)*. This system of 12 regional banks, supervised by the Federal Home Loan Bank Board, provided short-term credit to member savings and loan associations. As stated by Gissler and Narajabad (2017), in 1934, the *Federal Savings and Loan Insurance Corporation (FSLIC)* began offering depositors in both federal- and state-chartered associations insurance against the loss of their savings.

In 1934, the Federal Housing Administration (FHA) was created as a U.S. government agency whose function was to insure mortgages, providing banks and other lending institutions with a guarantee that their housing loans would be adequately secured, and providing the housing industry with new financial stimulus during a time when almost no new homes were being built.

Because FHA-insured mortgages, on the homes as collateral for these mortgages, must conform to certain construction and mortgage contract standards, the agency has had great influence on the housing and mortgage-lending industries, as cited by the HUD (n.d.-d).

By 1937, the federal role in housing was further broadened under the *Housing Act of 1937*, establishing the *U.S. Housing Authority* (renamed the Federal Public Housing Administration in 1942 and the *Public Housing Administration* in 1947) as a permanent agency charged with building and subsidizing local low-income housing projects. This created the basic structure for the nation's system of public housing. According to Fish (1979), more than 1.5 million dwelling units have been erected under this act.

Gap Period (1937–1944: Focus on World War II)

The era of the 1930s was one of collapse and depression (1930–1933) and rescue and recovery (1933–1939). The construction industry was paralyzed; there were housing bond scandals, the disappearance of credit, and rampant foreclosures. The New Deal and Congress focused on creating jobs and housing, and rescuing middle-class homeowners and lenders through the FHA. At this time housing professionals and the real estate lobby emerged to fight over public housing and the Housing Act of 1937, as written by Keith (1973).

By World War II, the federal role in promoting housing construction and standards, and stable housing markets, had been established, and a burgeoning commitment to low-income housing was evolving.

During World War II, private housing production virtually stopped, as U.S. raw materials were diverted to the war. At the end of the war, extreme pent-up demand from over 15 years of marginal construction, and the needs from returning military personnel, created a severe shortage of housing of several million homes, according to Jacobs (1986) and Blount (2016). The Migration Policy Institute (2019) cited that from 1940 to 1945, population immigration and growth was impacting the Administration, Congress, and localities; population, immigration, and growth totaled over 200,000. According to Keith (1973), during this time there emerged local power centers through city halls, housing authorities, and labor unions.

Creation of National Housing/Loan Corporations (Postwar Recession)

In 1948, the Federal National Mortgage Association (FNMA or "Fannie Mae") was created and became a private (Government-Sponsored Entity – GSE) organization in 1968. This institution provides a secondary market (liquidity) for FHA and Veterans Administration (VA) home mortgages by standing ready to purchase such (standardized) loans from financial institutions.

The federal government responded to post-WWII urban concerns with the passage of the Housing Act of 1949, as well as through the Federal-Aid Highway Act of 1956, creating the national highway system. The former created the Urban Redevelopment Agency and gave it the authority to subsidize three fourths of the cost of local slum clearance and urban renewal. Under the act, "primarily residential" and "blighted" urban areas could be condemned, cleared of buildings, and sold for private redevelopment; property could be taken and condemned, cleared of buildings, and sold to private redevelopers under eminent domain procedures.

This legislation established the goal of providing decent homes for all U.S. households, according to Fish (1979). The American Planning Association (2014) cited that this act resulted in relocation problems for many minority and poverty-stricken communities, giving rise to much *criticism of urban renewal*, particularly where luxury housing or office/retail buildings were built on the cleared land.

According to the U.S. Department of Veterans Affairs (n.d.), the mortgage insurance programs of the FHA, and the guaranteed home loans provided by the Veteran's Administration, helped reduce the housing shortage and improve the quality of housing; *new construction between 1945 and 1955* exceeded the levels of home building achieved between *1915 and 1925*, with the majority of these new dwelling units being single family in the suburbs. This suburban development was heavily influenced by FHA's subdivision guidelines, as written by Michel and Ligon (2015).

Effects of World War II (Recession followed by Prosperity)

The 1940s started out being distracted by demobilization, the veterans' housing crisis, and a premature attempt to industrialize the housing industry. The public at this time expressed revulsion against war-time controls and the political spectrum swung to the Republicans in 1946. *This crystallized the right-wing opposition to housing programs* and the emergence of new Democratic leadership in Congress. The Congressional Record from the Library of Congress (n.d.) stated that during this time there was a rural-urban split in Congress and established the role of mayors, labor, and Northern Negros in the housing policy debate.

The struggle over the Housing Act of 1949 was largely due to the political impact of Harry Truman's (Democrat) upset victory in 1948 and the strength of the real estate lobby (NAR). There was demagoguery by opponents of housing legislation on racial segregation issues. The proponents from public interest lobbies and the President struck against the real estate lobby for final victory, according to Betters (1949).

Post-World War II Era (1953–1954 and 1957–1958: Recessions)

The 1950s were when the first steps to meet the housing crisis in cities really took hold. This was a time of establishing urban coalitions like the Urban League (UL) and National Association for the Advancement of Colored People (NAACP) for housing, while the real estate lobby (NAR) was exploiting neighborhood racial biases to block public housing locally, as cited by Gross (2017).

According to Fish (1979), the Housing Act of 1954 modified urban plans and renewal activities by requiring communities to adopt code enforcement, relocation, and other measures that would prevent further spread of urban blight. It also established new FHA mortgage insurance programs to help relocate slum residents and to encourage new urban construction.

The 1960s was primarily a period of civil rights. Acts like the Civil Rights Act of 1964, Voting Rights Act of 1965, and Civil Rights Act of 1968 all assisted minorities in being able to freely participate in the capitalist system free of discrimination. According to the National Archives (2018), this included the housing industry, ensuring that discrimination against minorities could not be accomplished with housing.

As referenced from Keith (1973), during the 1960s, under President Johnson's War on Poverty, postwar federal involvement in housing had grown so large and complex that it became necessary to create a new federal body, the Department of Housing and Urban Development (HUD), to administer the various housing programs. The time period from 1953 to 1957 was one of gradualism and stalemate. President Eisenhower was trying to seek a Republican-business consensus on housing. This was a time of triumphs for the real estate lobby and is reflected in the moderate proposals put forward in the Housing Act of 1954 and the growing participation of private enterprise in housing.

Gap Period (1954–1961: Recession and Divided Government)

From 1957 to 1961, Eisenhower moved to the right and the monetary climate was one of restraint and tight money, causing a recession in the home-building industry. The Democrats were swept into power in 1958, thus having great implications for housing, although Eisenhower vetoed housing legislation. From 1961 to 1963, there seemed to be more promise and direction. As written by John F. Kennedy Presidential Library and Museum (2018), the Housing Act of 1961 was the first major appeal to the cities and the lower middle class regarding housing issues.

The Housing Act of 1964 liberalized FHA procedures to speed the processing of FHA-insured mortgages, setting up a low-interest housing-rehabilitation loan program, and land provision for public or nonprofit housing, according to Fish (1979). The Housing Act of 1965 established the Cabinet-level Department of Housing and Urban Development (HUD), formed from the Housing and Home Finance Agency (HHFA), and incorporated the Federal Housing Administration, the Federal National Mortgage Association, the Public Housing Administration (PHA), the Urban Renewal Administration (URA), and the Community Facilities Administration (CFA), as mentioned by HUD (2014) and Mitchell (1985).

The Demonstration Cities and Metropolitan Development Act of 1966 created the Model Cities Program which sought to solve housing, education, employment, welfare, and health problems of slum dwellers. The garden city movement that preceded the program began in the United Kingdom by Ebenezer Howard. This movement of urban planning later spread to the United States in major cities like Chicago under Daniel Burnham. According to the International Garden Cities Institute (n.d.), the program sought to revitalize urban communities through coordinating health, recreation, education, welfare, housing, and employment programs.

Great Society (1960–1961: Recession)

As cited by Keith (1973) and Rohn (2014), from 1963 to 1966, Great Society programs were shadowed by the Vietnam War and transition from J.F. Kennedy to Lyndon Johnson. The Great Society programs were aggressive programs confronting civil rights, the suburbs, and their impact on housing. There were greater promises to the poor, cities, housing and community development. These programs were eventually faced with the financial dilemma of the Vietnam War, which cost about $168 billion, or $950 billion in current dollars.

As written by Fish (1979), a principal objective of the Housing and Urban Development Act of 1968 was to encourage the creation of "New Communities." But developers did not have sufficient financial resources for projects of such size. Under the act, the federal government undertook to guarantee bonds marketed to raise funds for these new communities. In return, the developers were to agree to certain economic and planning standards, providing housing for nearly all income groups and family sizes.

This act set up new programs allowing FHA to insure mortgages for houses and apartment buildings on which interest was paid by the occupant or developer, as low as one percent (1%), and a Rent Supplement program, according to National Low Income Housing Coalition (2016).

Rise of New Conservatism (1969–1976: Era of Stagflation)

The FNMA and the Government National Mortgage Association ("Ginnie Mae"), created in 1968 as a counterpart to "Fannie Mae," assist the housing industry by *purchasing FHA-VA mortgages* whenever lenders need funds; and was the first to issue Residential Mortgage-Backed Securities (RMBS).

As cited by Ginnie Mae (n.d.) and Jacobs (1986), due to the Government-Sponsored Entities (GSEs) guaranteeing the coupon bond interest rates, as a credit enhancement, reducing credit default risk, lowered overall mortgage interest rates – thereby encouraging lenders to originate more residential mortgages, and provide mortgage institutions with a means to expand their lending power. The association is still around today and is the only institution of its type to follow the "full faith and credit" clause of the U.S. Constitution.

In the 1960s, housing needs of the elderly and those with moderate incomes were added as target groups for housing subsidies. Attempts to improve housing for racial minorities culminated with the passage of the 1968 Civil Rights Act and subsequent Supreme Court decisions (Brown vs. Board of Education). These actions have substantially reduced, but by no means eliminated, racial discrimination in housing markets, as mentioned by HUD (n.d.-a).

The 1966–1968 period was one of urban crisis. The war forced the home-building industry into recession. Congress took the initiative to build 26 million homes in ten years under the 1968 Act, cited by HUD (n.d.-a). According to Massey (2015), the Housing and Urban Development Act of 1970 authorized greater *outlays* for housing subsidy programs and additional funds for rent supplements to moderate-income households up to

$500 million: created a Community Development Corporation to encourage "new towns."

The Emergency Home Finance Act of 1970 authorized the Home Loan Bank System to reduce interest rates on home mortgages by means of a federal subsidy, establishing mechanisms for a secondary market in conventional mortgages through the creation of Federal Home Loan Mortgage Corporation (FHLMC).

As stated by Finegold, Schardin, and Wherry (2004), finally, during the Nixon era of the early 1970s, the U.S. government instituted several block grants. Although his proposal for them was rejected by Congress, which was controlled by the Democratic Party, President Ford would later pass them several years later. These grants essentially provided the states with some discretion as to what to spend the money on. As such, states now had some capital to invest in housing projects for their residents.

Gap Period (1974–1986: Divided Government)

The Housing and Community Development Act of 1974 sharply reduced new construction subsidies and shifted the emphasis for low-income housing to the use of (Section 8) leased-housing allowances and vouchers on rental units, and block grants to states for housing. The HUD Exchange (1999) reported that although there have been several shifts in policy since 1974 – most recently there has been a severe curtailment of federal programs to meet housing goals set out in the 1974 act; however, in a historical context the broad outlines of the legislation remain intact.

The 1970s reflected an era of uncertainty under the Nixon Administration and the eventual dissatisfaction of U.S. housing policy among mayors, labor leaders, and liberal organizations, as mentioned by Keith (1973). The housing affordability index between 1970 and 1980 dropped from 160 in 1970 to 80 in 1980, which is a 50% decline; and the median existing home price rose from $23,000 to $62,200, up 10.5% per year, well above the annual inflation rate of 8.6%, indicating housing wealth accumulation for (upper-class) home owners (U.S.A., 2018).

Era of Renewed Conservatism (1980, 1981–1982, 1990–1991: Recessions)

By the 1980s, federal spending on housing was dramatically reduced under President Reagan; the budget for public housing and
subsidized rentals dropped from $28 billion in 1980 to $9 billion
in 1989, down 68%, despite an increase in poverty. From 1980 to
1989, HUD was wracked by political favoritism, fraud, and mis-
use of funds. Trends reversed starting in 1989 under HUD secretary Jack Kemp, leading to 1990 legislation funding construction,
rehabilitation, and tenant ownership of public housing, as stated
by Bratt and Keating (1993).

Policy Gap (1990–2001: Divided Government)

Starting in the 1980s, because of decreased federal funds, many
U.S. cities were forced to increase their roles in providing housing. For example, from 1978 to 1983, HUD's budget dropped from
$83 billion to $18 billion, down 78.3% over the five-year period, or
26% per year, reported by the Western Regional Advocacy Project (n.d.).

Among the provisions of the National Affordable Housing Act
of 1990 are funds to help tenants of public housing buy their own
units, increasing their pride of ownership as well as their economic prospects. Funded Homeownership and Opportunity for
People Everywhere (HOPE I) program to help poor residents buy
their public housing units, as mentioned by American Legislative
Exchange Council (1995). According to the U.S. Department of
Housing and Urban Development (n.d.), this act also allows the
federal government to authorize block grants to state and local
governments for construction and rehabilitation of public housing.

Moreover, in 1992, the HOPE VI program began, which essentially worked as a demolition and replacement program that
sought to improve old homes for poorer residents. The program
was officially created by law in 1998 after President Clinton and
HUD Secretary Henry Cisneros revitalized the program, as cited
by HUD (n.d.-c).

Sharp Shift in Housing Policy
(2001 and 2007–2009 Recession)

Policy Gap (2003–2009: War on Terrorism)

As mentioned by Jan, Dewey, and Stein (2018), throughout the 21st century, there has been a slight shift toward more demand-side legislation. While much of this resulted from the 2008–2009 financial crisis, leading to the resurgence of Keynesian economics, this has even continued during the Obama administration.

Legislation like the Housing and Economic Recovery Act of 2008 have all been attempts to combat the 2008–2009 recession, and resulted in the government's greater role in the economy, as written by Cassidy (2017). The government takeover of Fannie Mae and Freddie Mac was an extreme example of government power, even though Republicans are currently attempting to privatize them once again, according to Stewart (2017). Even so, the supply-side legislation bias continues into the modern era, and even with a slight shift toward demand-side legislation, this bias is likely to continue into the future.

Policy Gap (2012–2017: Divided Government)

The second noticeable trend of the 21st century, in addition to supply-side housing policy bias, has been skyrocketing housing prices beginning around the time following the 2008–2009 recession, and becoming exceedingly noticeable by 2018, stated by Lerner (2018). The home affordability index dropped from its peak of 204 in 2012 to 161 in 2017, down 21%, or 5% per year, to the lowest level since 2008; and home prices rose from a median of $221,700 in 2008 to $315,200 in 2017, up 42%, or approximately 5% per year during the same trend period, reported by ATTOM Data Solutions (2018).

From 2018 to 2020, there were no housing policies initiated from the executive branch or congress; however, the Biden-Harris Administration along with democratic party control of both the house and the senate from 2021 to 2022, and potentially 2023 to 2024, increases the probability of housing policy legislation to address the current housing shortage and lack of affordability.

In the previous section, a history of U.S. institutional housing policy and political analysis was conducted, leading to the conclusion that there is supply-side housing policy bias; the next section of this research investigates economic factors and policy tools driving supply and demand for housing.

Citations

American Legislative Exchange Council. (1995, January 4). *The homeownership and opportunity for people everywhere act.* The Heartland Institute. https://www.heartland.org/publications-resources/publications/the-homeownership-and-opportunity-for-people-everywhere-act

American Planning Association. (n.d.). *Housing act of 1949.* Retrieved 2018, from https://www.planning.org/awards/2014/1949housingact.htm

ATTOM Staff. (2018, May 21). *Americans departing high-priced housing markets for more affordable alternatives.* ATTOM Data Solutions. https://www.attomdata.com/news/market-trends/home-sales-prices/home-affordability-top-departure-and-arrival-markets/.success

Betters, P. V. (1949), The United States housing act 1949. *Annals of Public and Cooperative Economics, 20*(3), 375–378. https://doi.org/10.1111/j.1467-8292.1949.tb01586.x

Blount, J. (2016). *Housing, not unemployment, major concern when World War II veterans returned home.* Historical Collection at the Lane. Hamilton Journal-News. https://sites.google.com/a/lanepl.org/columns-by-jim-blount/2016-articles/housing-not-unemployment-major-concern-when-world-war-ii-veterans-returned-home

Bratt, R. G., & Keating, W. D. (1993). Federal housing policy and Hud: Past problems and future prospects of a Beleaguered Bureaucracy. *Urban Affairs Quarterly, 29*(1), 3–27. https://doi.org/10.1177/004208169302900101

Cassidy, J. (2017, January 9). Obama's economic record: An assessment. *The New Yorker.* https://www.newyorker.com/news/john-cassidy/obamas-economic-record-an-assessment

Finegold, K., Schardin, S., & Wherry, L. (2004, April 21). *Block grants: Historical overview and lessons learned.* The Urban Institute. New Federalism: Issues and Options for States, Series A, (A-63). http://webarchive.urban.org/UploadedPDF/310991_A-63.pdf

Fish, G. S. (1979). *The story of housing.* Macmillan Publishing.

Ginnie Mae. (n.d.). *Our history.* Retrieved 2018, from https://www.ginniemae.gov/about_us/who_we_are/Pages/our_history.aspx

Gissler, S., & Narajabad, B. (2017, October 28). *The increased role of the federal home loan bank system in funding markets, part 1: Background.* Board of Governors of the Federal Reserve System. https://www.federalreserve.gov/econres/notes/feds-notes/the-increased-role-of-the-federal

Gross, T. (2017, May 3). *A 'forgotten history' of how the U.S. government segregated America.* NPR. https://www.npr.org/2017/05/03/526655831/a-forgotten-history-of-how-the-u-s-government-segregated-america

HUD Exchange. (1999, January). *Housing and community development act of 1974.* https://www.hudexchange.info/resource/2184/housing-and-community-development-hcd-act-of-1974/

Jacobs, B. G. (1986). *Guide to federal housing programs.* Bureau of National Affairs.

Jan, T., Dewey, C., & Stein, J. (2018, April 25). HUD Secretary Ben Carson to propose raising rent for low-income Americans receiving federal housing subsidies. *The Washington Post.*-home-loan-bank-system-in-funding-markets-part-1-background-20171018.htm

John F. Kennedy Presidential Library and Museum. (2018). *Legislative summary: Housing.* Retrieved 2018, from https://www.jfklibrary.org/Research/Research-Aids/Ready-Reference/Legislative-Summary-Main-Page/Housing.aspx

Keith, N. S. (1973). *Politics and the housing crisis since 1930.* Universe Books.

Markels, A. (2008, February 28). *Comparing today's housing crisis with the 1930s.* U.S. News & World Report. https://money.usnews.com/money/personal-finance/real-estate/articles/2008/02/28/comparing-todays-housing-crisis-with-the-1930s

Massey, D. S. (2015). The legacy of the 1968 fair housing act. *Sociological Forum, 30*(S1), 571–588. https://doi.org/10.1111/socf.12178

Michel, N., & Ligon, J. (2015, May 11). *The federal housing administration: What record of success?.* The Heritage Foundation. https://www.heritage.org/housing/report/the-federal-housing-administration-what-record-

Migration Policy Institute. (2019). *Legal immigration to the United States, 1820-present.* https://www.migrationpolicy.org/programs/data-hub/charts/Annual-Number-of-US-Legal-Permanent-Residents

Mitchell, J. P. (Ed.). (1985). *Federal housing policy and programs: Past and present.* Routledge.

National Archives. (2018, April 25). *The Civil Rights Act of 1964 and the equal employment opportunity commission.* https://www.archives.gov/education/lessons/civil-rights-act

National Low Income Housing Coalition. (2016, May 3). *State Rent Supplement Program.* Retrieved 2018, from http://nlihc.org/oor

Nigro, C. (2018, June 7). *Tenement homes: The outsized legacy of New York's notoriously cramped apartments.* New York Public Library. https://www.nypl.org/blog/2018/06/07/tenement-homes-new-york-history-cramped-apartments

Rohn, A. (2014, January 22). *How much did the Vietnam War cost?* The Vietnam War. https://thevietnamwar.info/how-much-vietnam-war-cost/

Social Welfare History Project. (2018). Tenement house reform. *Social Welfare History Project.* http://socialwelfare.library.vcu.edu/issues/poverty/tenement-house-reform/

Stewart, E. (2017, July 19). *Fannie Mae and Freddie Mac would be privatized under proposed House budget.* https://www.thestreet.com/story/14233125/1/fannie-mae-and-freddie-mac-would-be-privatized-under-proposed-house-budget.html

The International Garden Cities Institute. (n.d.). *A prospectus for the international garden cities institute.* https://www.gardencitiesinstitute.com/sites/default/files/documents/igci_prospectus_low_res.pdf

U.S.A. (2018). Federal home loan mortgage association (Freddie Mac). http://www.freddiemac.com/

U.S. Department of Housing and Urban Development. (n.d.). *The Federal Housing Administration (FHA).* Retrieved 2018, from https://www.hud.gov/program_offices/housing/fhahistory

Western Regional Advocacy Office. (n.d.) *History of slashing HUD budget sources.* http://weap.org/uploads/fact%20sheets/SourcesWRAP-HistoryofSlashingHUDBudgetFactSheet-Final.pdf

4 U.S. Housing Economics

Factors Driving Demand and Supply for Housing

Demand Factors

Housing demand is a preference for housing services subject to constraints of income and price. The household is the basic unit of housing demand, and as defined by the U.S. Census, a household is the person or group of people who jointly occupy a dwelling unit and who constitute a single economic unit, as reported by the U.S. Census Bureau (n.d.). Determinants of household formations are the number of total households, economic and demographic characteristics, tastes and preferences, prices of substitute and complementary goods, and expectations, according to Carn et al. (1988).

The passage of the baby boomers (81.3 million), Gen X (65.6 million), and the emerging Echo Boomers/Millennials (Gen Y: 83.1 million) will be especially important in determining the number and type of housing units required over the next 25 years, because household formations and demand for second homes vary systematically with the age of individuals, as cited by CNN (2017). In 2017, there were about 136.57 million housing units in the United States. Based on previous trends, this number could rise by at least 30 million new units. As mentioned by Statista (2017), since there are more Millennials than Gen Xers, the number probably would be more than the predicted 30 million.

DOI: 10.1201/9781003223436-4

As cited by Marcin and Kokus (1975), differences in housing demand remain fairly stable over long periods of time due to regular patterns of income, income expectations, assets, family status, and preferences of the life cycle. Demographic factors affecting housing demand are rate of household formations, size and age mix of population, and household size. Other factors affecting housing demand are inflation and deterioration.

As mentioned by Hanushek and Quigley (1980), the demand for housing is similar to any other commodity. The demand function is negative; therefore, if home prices rise rapidly, demand at some point declines. The price elasticity of housing is affected by tax treatments, interest rates, cost of construction and maintenance and repairs. If any of these cost factors drive up the price of homes, demand is affected negatively, as stated by Harrington (1989). Other factors important to determining demand for housing services are permanent income, marginal tax rate, income growth, expected rate of return on housing, family size, and location of home, as cited by Haurin and Lee (1989).

According to Abraham and Hendershott (1996), increasing demand for housing is also reflected in rates of home price appreciation. Factors contributing to home price appreciation rates are cost inflation, employment growth, real income growth, and change in interest rates. Housing demand is the quantity of housing desired by households individually and in aggregate. Housing demand is negatively related to the price of housing, and positively related to household income. As the price of housing falls due to lower interest rates, construction costs, or real tax burdens, the demand for housing, or household formations, increases, as written by Pozdena (1988).

These demand-side factors, such as population, job growth, home price appreciation, etc., help to target demand-side policy tools.

Demand-Side Policy Tools

The main demand-side policy tools being used are demand-side subsidies, housing allowances, and vouchers.

As cited by Vliet (1998), during the late 1970s, rapidly rising housing costs due to hyperinflation replaced poor conditions as the major housing problem, and vouchers were considered less expensive than building new units. During the 1980s and 1990s, under the Republican administrations, federal government involvement in housing production virtually stopped.

Demand-side subsidies go to housing consumers and lower the cost of consuming housing services. Subsidies make it more affordable for households to purchase or rent a home. Subsidies come in many forms: direct cash payments, housing allowances or vouchers, or preferential homeownership tax treatment. In the United States, the largest subsidy comes in the form of preferential tax treatment of mortgage interest payments, as opposed to Section 8 vouchers.

Housing allowances involve cash payments to households, but in many cases, money is paid directly to landlords. The value of the voucher is directly related to local rent levels, household size, and income. Payments are restricted and determined by maximum income levels.

The acceptance and use of vouchers in America reflect the perception that housing problems are a question of affordability. By giving vouchers to low-income households, they receive more buying power and are able to meet housing costs and compete for housing in the market. This policy tool is favored over other forms of subsidies because it allows for targeting specific socio-economic groups; therefore, it is superior to production subsidies. However, these vouchers may also push rents higher in a given market due to rising effective demand, thereby benefiting housing suppliers and developers without addressing the problem of housing shortages.

Preferential tax treatments for homeowners come in the form of paying less tax through mortgage interest deductions. Tax expenditures are reduced through tax deductibility of interest payments on home mortgages; capital gain shields of up to $250,000 for single persons and up to $500,000 for married couples, and up to $750,000 in total mortgage interest (IRS, n.d.).

This tax policy is inequitable in the sense that it benefits mostly those households with high incomes (+$100,000) and expensive houses, is inefficient in allocating scarce resources, increases the value of homes above their economic value, and distorts housing choice. It is therefore the largest middle-class subsidy. This policy tool is fiscal and political in nature and not necessarily a good housing policy tool, as reported by McManus (2015) for MarketWatch.

Housing subsidies in the form of vouchers and allowances and reduction in tax expenditures are appropriate policy tools in periods of slack demand, and should be used only when increasing demand brings about benefits without creating economic inefficiencies, distortions, or social injustices.

Demand-side factors also drive supply-side factors and policy tools.

Supply Factors

According to Smith (1970), the main supply-side factors driving production (housing construction) and affordability are level of historical housing starts, rate of household formations, rate of removals, number of vacant units, rent growth, and interest rates.

Housing supply can be thought of as stock of units, flow of services, or quality of accommodations from the stock. The U.S. Census Bureau (2018) estimates that there are over 137 million dwelling units in the United States as of 2017.

Housing supply is not easy to determine due to the difficulty of assessing and measuring the supply of available land for development. But it is assumed that in the long run housing supply is infinite, and that developers will find a higher and better use of the land, and if that means building up or tearing down an existing structure, they will.

As written by Weber and Devaney (1996), rapid home price appreciation, above construction costs, gives developers incentive to produce more housing, but before developers will develop, they must be convinced that consumers are confident enough in the current employment and economic environment. The market for new homes depends partly on consumer attitudes and perceptions embodied in consumer confidence indices.

The true relationship between private housing starts and the economy is a function of the money supply (M1/M2) and overall performance of national output measured by Gross Domestic Product (GDP). Historically, it has been shown that growth of deposits at savings and loans was a significant determinant of housing starts in 1966 and 1971, but not in other years, as mentioned by Johnson (1982).

These supply-side factors help to target supply-side policy tools.

Supply-Side Policy Tools

Major supply-side policy tools driving housing production and affordability are subsidies to lower the cost of building homes, assistance in the actual construction of homes, and overall support for those building homes.

Also cited by Vliet (1998), supply-side subsidies go to suppliers of housing and reduce the costs of provision. These subsidies reduce the cost of providing housing by builders and landlords. Subsidies can take the form of direct cash payments, fiscal concessions on taxes, reductions in loan payments, subsidized mortgage interest rates, loan underwriting warranties, or land grants for development.

Supply-side programs in the United States are limited in nature. Supply-side programs seek to increase the supply of housing through construction or support for housing providers. The 1937 and 1956 Housing Acts provided for government involvement in slum clearance and construction of affordable housing units.

Two U.S. government supply-side programs are Section 202 and Section 221(d)(3). These programs are designed to provide new housing for seniors through nonprofit community development corporations. By the 1990s, production of new Section 202 projects had dwindled to 6,000 units per year, compared to 20,000 units in the late 1970s, down over 70%. As of 2019, roughly 400,000 low-income elderly households have been produced (Couch, 2020; HUD User, 2020).

The principal purpose of supply-side housing policies is to reduce housing shortages, reduce unemployment, and promote economic growth. Over the years policymakers have been criticized by the large-scale poorly planned housing developments produced in the public sector, reducing support for supply-side programs. Supply-side programs allow the government to influence the design, location, occupancy, and rent levels of housing. Critics see government involvement as lacking in degree of decentralized control and individual choice; therefore, they advocate demand-side subsidies as a preferred alternative.

The next section of this research discusses the methodology, analytical procedures, and results of the historical policy and econometric analysis.

Citations

Abraham, J. M., & Hendershott, P. H. (1996). Bubbles in metropolitan housing markets. *Journal of Housing Research, 7*(2), 191–207.

Carn, N., Rabianski, J., Racster, R., & Seldin, M. (1988). *Real estate market analysis: Techniques and applications.* Prentice Hall.

CNN Editorial Research. (2017, August 27). *American generation fast facts.* CNN. Retrieved 2017, August 27 from https://www.cnn.com/2013/11/06/us/baby-boomer-generation-fast-facts/index.html

Couch, L. M. (2020, Summer). *Falling short: Federal housing assistance is failing older adults.* Generations American Society on Aging. https://generations.asaging.org/federal-housing-assistance-older-adults-failing

Hanushek, E. A., & Quigley, J. M. (1980, February 22). What is the price elasticity of house demand? *Review of Economics and Statics, 62*(3), 449–454.

Harrington, D. E. (1989). An intertemporal model of housing demand: Implications for the price elasticity. *Journal of Urban Economics, 25*(2), 230–246.

Haurin, D. R., & Lee, K. (1989, November). A structural model of the demand for owner-occupied housing. *Journal of Urban Economics, 26*(3), 34–360.

IRS. (n.d.). *Credits and deductions for individuals.* Retrieved 2018, from https://www.irs.gov/credits-deductions-for-individuals

Johnson, B. (1982). *Resolving the housing crisis: Government policy, decontrol, and the public interest.* Pacific Institute for Public Policy Research.

Marcin, T. C., & Kokus, J. (1975). Economic and demographic factors affecting demand into the 1980s and 1990s. *Real Estate Economics, 3*(3). 81–93.

McManus, J. O. (2015, December 25). *5 ways to protect your estate from capital gain taxes.* MarketWatch. https://www.marketwatch.com/story/5-ways-to-protect-your-estate-from-capital-gains-taxes-2015-12-25

Pozdena, R. J. (1988). *The modern economics of housing: A guide to theory and policy for finance and real estate professionals.* Quorum Books

Smith, W. F. (1970). *Housing: The social and economic elements.* University of California Press.

Statista Research Department. (2017). *Number of housing units in the United States from 1975 to 2017 (in millions).* Statista. Retrieved 2017, from https://www.statista.com/statistics/240267/number-of-housing-units-in-the-united-states/

U.S. Census Bureau. (2018, May 24). *Census Bureau reveals fastest-growing large cities.* https://www.census.gov/newsroom/press-releases/2018/estimates-cities.html

U.S. Census Bureau. (n.d.) *Subject definitions.* https://www.census.gov/programs-surveys/cps/technical-documentation/subject-definitions.html#householdfamily

U.S. HUD. (2020). *Picture of subsidized households.* https://www.huduser.gov/portal/datasets/assthsg.html

Van Vliet, W. (Ed.). (1998). *The encyclopedia of housing.* Sage Publications, Inc.

Weber, W., & Devaney, M. (1996). Can consumer sentiment surveys forecast housing starts? *Appraisal Journal, 4*, 343–350.

5 Research Questions and Hypotheses

Housing Policy Programs, Politics, and Economics

Based on a review of the economic literature and housing policy history, this chapter of the study tries to answer basic research questions relating to U.S. urban housing policy and factors driving supply and demand for housing. Answers to these questions put U.S. urban housing policy in a historical context; and reveals legislative, psychological, economic, demographic, financial, and institutional drivers of housing supply and demand.

Basic Research Questions

Urban Housing Policy Questions

- Why did housing policy in the United States arise?
- What were the political dynamics surrounding these policies?
- Was there some immediate crisis widely perceived by the general public?
- What agency (agencies) is (are) charged with administering the program?
- Was/is information available about the policy and its impacts on the general public?
- If crisis played a role in initiating policy, was policy changed when crisis abated?

DOI: 10.1201/9781003223436-5

Urban Housing Economic Questions

Supply and Demand

- What are the legislative policy (Monetary/Fiscal) drivers of housing supply and demand?
- What are the economic (GDP/Employment/Income) drivers of housing supply and demand?
- What are the psychological (Consumer and Business Confidence) drivers of housing supply and demand?
- What are the market (Home Prices/Affordability) drivers of housing supply and demand?
- What are the demographic (Households/Population) drivers of housing supply and demand?
- What are the financial (Money Supply/Interest Rates) drivers of housing supply and demand?
- What are the institutional (Financial/Lending System) drivers of housing supply and demand?

Hypotheses

Hypotheses tests of data specifically address urban housing policy issues and the economics of housing supply and demand. Limits to the number of hypotheses tested are determined by the scope of research, data availability, and analysis plan. These hypotheses are tested for significance using observational techniques, correlation analysis, and Ordinary Least Squares (OLS) regression analysis. A comprehensive list of hypotheses is given below; the actual number hypotheses tested is limited based on time, data availability, and scope of the research.

Urban Housing Policy Hypotheses

H1: Housing policy grew out of housing crises.
H2: Housing policy is directly related to party politics.
H3: Lack of quality housing and rising health concerns lead to legislative action.

Urban Housing Economic Hypotheses

Demand

H1: Mortgage interest rates drive housing demand.

H2: Housing demand is positively correlated with Gross Domestic Product (GDP).

H3: Housing demand is positively/negatively correlated with employment growth/unemployment rate.

H4: Housing demand is positively correlated with consumer confidence.

H5: Housing demand is negatively correlated with home price appreciation, and positively correlated with housing affordability.

H6: Housing demand is positively correlated with growth in homebuyer demographic cohorts (22–55 years old).

H7: Housing demand is negatively correlated with the number of persons per household.

H8: Housing demand is positively correlated with income growth rates.

Supply

H1: Housing supply is negatively correlated with mortgage interest rates.

H2: Housing supply is positively correlated with Gross Domestic Product (GDP).

H3: Housing supply is positively correlated with consumer and business confidence.

H4: Housing supply is positively correlated with employment growth, and negatively correlated with unemployment rates.

H5: Housing supply is positively correlated with home price appreciation.

H6: Housing supply is positively correlated with growth in homebuyer demographic cohorts.

6 Methodology

History U.S. Housing Policy, Factor Analysis, and Forecasting

The methodology used in this section of the study is separated into two types: (1) historical analysis and (2) quantitative analysis.

Historical analysis is the collection, chronology, and categorization of significant historical policy and political events. Historical events have impacted supply and demand for housing. Historical analysis applied in this section takes federal housing legislative events and categorizes them over time as either supply-side or demand-side oriented.

This categorization makes the case for housing policy bias over time. Quantitative analysis takes collected data and applies statistical techniques to assess the relationship between dependent variables and independent variables. Depending on the results of the statistical analysis, and prior research in the field of study, a forecast model is constructed.

Data Collection

Data collection for this section involves secondary sources. Sources of information are mainly government agencies (websites), private research institutions, public and private universities, and industry groups. Sources of information (data) used in the data and regression analysis come from:

- ATTOM Data Solutions
- Columbia University Center for Business Cycle Research
- Congressional Record

DOI: 10.1201/9781003223436-6

- Construction Industry Board
- Federal Reserve Board of St. Louis
- Federal Reserve System
- National Association of Realtors
- U.S. Department of Commerce, Bureau of the Census
- U.S. Department of Housing and Urban Development (HUD)
- U.S. Department of Labor
- U.S. Department of Veterans Affairs
- Etc.

Analytical Procedures

The following are the methodological phases of the research:

Phase I: Collect, assemble, and classify significant U.S. housing policy and politics surrounding legislation over time.

Phase II: Conduct literature review to determine significant variables that determine supply and demand for housing in the United States.

Phase III: Collect housing supply and demand data from secondary sources.

Phase IV: Calculate and transpose data. Calculate year-over-year percent changes and four quarter moving averages for time series.

Phase V: Test the degree of relationship and statistical significance between the dependent variable and independent variables using correlation analysis.

Phase VI: Test the degree of relationship and statistical significance between the dependent variable and independent variables using regression analysis.

Phase VII: Test for multicollinearity between independent variables using correlation analysis.

Phase VIII: Eliminate independent variables with high correlation to each other, and retain variables with statistical and theoretical significance.

Phase IX: Re-test the degree of relationship and statistical significance between the dependent variable and independent variables using regression analysis for forecast model development.

Phase X: Back-test forecast model by inputting actual historical data into forecast models, and graph the results.

Phase XI: Once the back-test is complete, input forecast independent variables into the forecast model to calculate projected supply- and demand-dependent variables.

Phase XII: Once the forecast-dependent variables have been produced, net the projected supply against the projected demand to arrive at the future housing gap.

Phase XIII: Housing gap is then graphed to determine long-run supply and demand conditions.

Phase XIV: Conclusions and recommendations are then made based on model results.

7   Data Analysis Plan

Quantitative-Qualitative
Statistical Analysis

Qualitative analysis used in this research involved searching the Congressional Record for all housing policy since 1930, when housing policy in the United States began under the New Deal. Then, the political parties in power were found, also using the Congressional Record, and divided by the time periods and by President, Senate, and House of Representatives. Finally, this information was analyzed by looking at which party enacted what policies, and when.

Quantitative analysis used in this research is mainly regression Ordinary Least Squares (OLS) analysis, coupled with correlation analysis and analysis of variance techniques. Data will be displayed graphically and in tables for easy identification of trends and relationships between the dependent variable and independent variables once the forecast models are determined, projections are conducted, and the supply-demand gap analysis is completed.

Qualitative Analysis

Content Analysis

In order to understand U.S. housing policy, the Congressional Record was utilized to find all housing policy within American since 1930. Year 1930 was soon after the start of the Great Depression, and soon before President Roosevelt began the New Deal, which essentially invented housing policy in the United States.

DOI: 10.1201/9781003223436-7

As such, the Congressional Record was searched over the past 88 years, from 1930 to 2018. After finding the policies, the 88 years were broken up into decades, and the policies were placed in their respective decades. The decades were then divided up into two-year election periods, and those periods without any policies were labeled as policy gaps. Those periods of time with a certain relationship between policies, such as New Deal policies, were labeled as policy blocks.

Then, the policies were divided between supply-side policies and demand-side policies in order to demonstrate the extreme supply-side policy bias. The party in power during each two-year period was also found from the Congressional Record and labeled according to policies enacted. The party in power was divided between the Presidency, Senate, and House of Representatives. From this analysis, we were able to determine policy bias, as well as policy bias by political party in office.

Quantitative Analysis

Single-Variable Regression Model

Single-variable regression is used to test the significance between dependent and independent variables individually.

The single-variable linear regression model is $Y_i = \beta_0 + \beta_1 X_i + u_i$
   where:

β_0 is the intercept. It is an average level of return of the independent variable if the X coefficient is zero.

β_1 is the explanatory X variable, or slope coefficient. It measures the rate of change in the conditional mean value of Y_i per unit change in X.

u_i is the stochastic, residual, or the random error term. The error term may represent the influence of those that are not explicitly in the model; even if we included all variables that determine Y levels, some "intrinsic randomness" is bound to occur that cannot be explained, no matter how hard we try.

Y_i is the dependent Y variable, the average response value for a given level of the X coefficient.

Multiple Regression

Multiple-variable regression is used to test the significance between dependent and independent variables in an overall model context. Final determination of the forecast model is based on "t," "F," "R-square," etc. diagnostic statistics, the theoretical inference.

$$E(Y_i) = \beta_1 + \beta_2 X_{2t} + \beta_3 X_{3t} + u_t$$

where β_2 and β_3 are known as partial regression or partial slope coefficients. Partial's meaning is as follows: β_2 measures the expected change in the mean value of Y, $E(Y)$, per unit change in X_2, holding the value of X_3 constant. The same rule applies for β_3.

R-Squared

Correlation analysis is used to determine the relationship of the model and variables in predicting the dependent variable. Correlation analysis was also used to screen independent variables and as a diagnostic tool to reduce multicollinearity in the model.

r^2, the coefficient of determination, measures the percentage of total variation in Y explained by the single explanatory (X) variable regression model.

R^2, the multiple coefficient of determination, measures the percentage of total variation explained by X_2 and X_3 jointly, in a multiple regression model.

ESS denotes explained (by the regression) sum of squares.

RSS denotes residual (or unexplained) sum of squares variation of Y values around the regression line.

TSS denotes total sum of squares; TSS=ESS+RSS.

The range of $r^2 = 0 <= r^2 <= 1$; 1 indicates a perfect fit and 0 indicates no relationship whatsoever.

If there is an insignificant variable (low t statistic) in the multiple regression model, and should be

$$1 = \frac{ESS}{TSS} + \frac{RSS}{TSS}$$

included because of economically intuitive reasons, the software will calculate a smaller coefficient for the insignificant "t" statistic. For example, with the employment and population variables highly significant and multifamily housing permits insignificant, the software calculates a regression equation: 7.16 + 5.24*EMPL + 12.3*POP +.00125*MPERM. In other words, the computer compensates in order to maintain a high correlation in the overall regression model with high F scores. The "F" ratio [(regression mean square)/(residual mean square)] should exceed the factor, i.e., score of four in order for the regression to be predictive.

8 U.S. Housing Policy Analysis and Results

Supply-Demand-Side Political and Policy Outcomes

Qualitative Analysis

U.S. government policies toward housing have been biased toward supply-side incentives. This supply-side bias is reflected in the history (categorization) of housing legislation. In almost every case of legislation, the policy bias has been toward increasing housing production. The only periods in which there is demand-side bias is during crises such as depressions and recessions.

A significant portion of demand-side legislation comes in the form of direct subsidies to low-income residents and grants to affordable housing developers. The majority of supply-side legislation takes the form of banking or finance (interest rate) subsidies or loan guarantees to developers and individuals.

Policy Analysis

Policy analysis was conducted, and based on the research, it has been found that policy is generally conducted during periods of recession or low economic output. For example, five supply-side policies were conducted during the Great Depression, while there was only a single policy during World War II, where economic output was high.

Supply-side legislation is meant to increase the supply of housing to increase competition and lower pricing. When more people can buy housing, the economy becomes stimulated as more houses are built, and the cycle continues, raising Gross Domestic

DOI: 10.1201/9781003223436-8

Product (GDP). In general, more housing policy is created when there is a Democratic Party administration or Congress.

Legislative Analysis

Legislative analysis was conducted, and based on the research, 28 supply-side policies have been conducted. They have been grouped by political party in power, date, and whether there had been a recession. The result is that policy is enacted as a frame predominantly during periods where the executive and legislative branches are controlled by Democrats. There are nine policy frames and six policy gaps.

Housing Policy Results

Supply-Side Legislation (Construction and Lending) – 28

Based on this research, 28 supply-side policies were identified from 1932 to 2018 (76 years). Supply-side legislation was categorized as those policies that improved housing quality, production, standards, affordability, capital flows, and incentives (taxes).

Housing Policy (1932–1946) – 15 years/7 policies

1   Federal Home Loan Bank System (1932)
2   Home Owners Loan Corporation (HOLC) (1933)
3   Resettlement Administration (1933)
4   National Housing Act (1934)
5   Housing Act (1937)
6   Servicemen's Readjustment Act (1944)
7   Farmers Home Administration (1946)

Policy Gap (1946–1948) – 3 years/0 policies

Housing Policy (1948–1954) – 7 years/3 policies

8   Federal National Mortgage Association (1948)
9   Housing Act (1949)
10 Housing Act (1954)

Policy Gap (1954–1960) – 7 years/0 policies

Housing Policy (1961–1974) – 9 years/8 policies

11 Housing Act (1961)
12 Housing Act (1964)
13 Housing and Urban Development Act (1965)
14 Demonstration Cities and Metropolitan Development Act (1966)
15 Housing and Urban Development Act (1968)
16 Housing and Urban Development Act (1970)
17 Emergency Home Finance Act (1970)
18 Equal Credit Opportunity Act (ECOA) (1974)

Policy Gap (1974–1986) – 13 years/0 policies

Housing Policy (1986–1990) – 5 years/2 policies

19 Tax Reform Act (1986)
20 National Affordable Housing Act (1990)

Policy Gap (1990–2000) – 11 years/0 policies

Housing Policy (2001–2003) – 3 years/3 policies

21 Economic Growth and Reconciliation Act (2001)
22 American Dream Downpayment Act (2003)
23 Jobs and Growth Tax Relief Reconciliation Act (JGTRRA) (2003)

Policy Gap (2003–2009) – 6 years/0 policies

Housing Policy (2009–2012) – 4 years/3 policies

24 Troubled Asset Relief Program (TARP) (2009)
25 Tax Relief, Unemployment Insurance Reauthorization, and Job Creation Act (TRUIRJCA) (2010)
26 American Taxpayer Relief Act (2012)

Policy Gap (2012–2016) – 5 years/0 policies

Housing Policy (2017–2018) – 2 years/2 policies

27   Tax Cuts and Jobs Act (TCJA) (2017)
28   Building a Better America Budget Plan (BBABP) (2018)

Note: from 2019 to 2020, there were no real supply-side housing policies initiated.

While demand-side legislation has been significantly less common than supply-side legislation throughout history, there has recently been a slight shift toward more demand-side legislation.

Demand-Side Legislation (Employment and Buying Power) – 7

Based on this research, seven (7) supply-side policies were identified from 1933 to 2018 (75 years). Demand-side legislation was categorized as those policies that provide subsidies to potential buyers, affect mortgage rates, and affect employment and wages.

Housing Policy (1933) – 1 year/1 policy

1   The Public Works Administration (1933)

Policy Gap (1933–1973) – 41 years/0 policies

Housing Policy (1974–1977) – 5 years/2 policies

2   Housing and Community Development Act (1974)
3   Housing and Community Development Act (1977)

Policy Gap (1977–2007) – 11 years/0 policies

Housing Policy (2008–2009) – 2 years/3 policies

4   Housing and Economic Recovery Act (HERA) (2008)
5   American Recovery and Reinvestment Act (ARRA) (2009)
6   Homeowners Affordability and Stability Plan (HASP) (2009)

Policy Gap (2009–2018) – 9 years/0 policies

Housing Policy (2018) – 1 year/1 policy

7 Making Affordable Housing Work Act (MAHWA) (2018)

Note: from 2019 to 2020, there were no real demand-side housing policies initiated.

U.S. Housing Supply- and Demand-Side Policy Bias

Table 8.1 shows the number of housing policies that have been supply side (28) vs. demand side (7). Clearly, there have been far more supply-side policies (80%) than demand-side policies (20%) in the history of the United States.

Even during periods of time such as the depression of the 1930s, where there was an extreme amount of housing policy, there were still far more supply-side policies than demand-side policies.

During that decade, supply-side policies accounted for 83% of all housing policy, while demand-side policies only accounted for 17%. Finally, whereas there was only one decade without

Table 8.1 U.S. Housing Supply- and Demand-Side Policy Bias

U.S. Housing Policy Democrats vs. Republicans

Date	# Supply	# Demand	% Supply	% Demand
1930–1940	5	1	83	17
1941–1950	4	0	100	0
1951–1960	1	0	100	0
1961–1970	7	0	100	0
1971–1980	1	2	0	100
1981–1990	2	0	100	0
1991–2000	0	0	0	0
2001–2010	5	3	63	37
2011–2018	3	1	75	25
Total	28	7	80	20

Source: Congressional Record
The bold numbers are to highlight the significance between the number in percentage of supply side policies compared to demand side policies

a supply-side policy, there were five decades without a single demand-side policy.

The prior section showed the history of housing policies, as well as the clear policy bias toward supply-side housing policies. The next section of this research shows a political analysis of the policies, demonstrating how generally *Democrats are far more likely to enact housing policy, supply-side and demand-side policies, than Republicans.*

Housing Policy and Political Scientific Analysis

As shown in the economic model forecast section, housing supply in the United States is expected to continue to outstrip housing demand, on the margin. Housing supply is expected to continue to exceed housing demand due to federal housing policy bias toward supply-side legislation, and the existence of government-backed mortgage institutions. Although during the 2008 recession, and even in the past few years, there seems to be a shift away from pure supply-side legislation, and an attempt at dismantling and privatizing government mortgage-backed institutions; however, the bias toward supply-side legislation, and its institutional framework, remains prominent.

Under the Trump administration, legislation proposed by HUD such as the Making Housing Affordable Work Act (MAHWA) aimed to fix the problem of rising housing prices through the demand side of the marketplace. This is a pattern in Republican administrations, as well as a general trend of the 21st century, even through the Democratic Obama administration, where demand-side housing legislations such as the American Recovery and Reinvestment Act (ARRA) and Housing and Accommodation Support Partnership (HASP) were passed.

The following section presents information regarding when housing policy was enacted and which political party was in power during these time periods. In addition, recession dates are noted.

Total demand-side policies: 7. Total supply-side policies: 28.

Policy Block Analysis

The data in Table 8.2 (see appendix) can be grouped into various periods of time defined by extreme output of policy followed by policy gaps, where no policy had been issued in several years.

Block 1: 1930–1950 – This is the first wave of policy innovation. During this time, there were eight policies under a Democratic president, and there were six policies under a Democratic president, House, and Senate.

Block 2: 1950–1960 – This is the Eisenhower era, where there are many policy gaps due to both the Korean War and divided government. However, because Eisenhower was a moderate Republican, this was a time where both parties worked together fairly well and so some policies were still enacted.

Block 3: 1960–1968 – This is the second wave of policy innovation, during both Kennedy's New Frontier and Johnson's Great Society. During this entire period, the Democrats controlled every branch of government.

Block 4: 1968–1972 – This is the Nixon era, where the executive branch was controlled by the Republicans, whereas the legislative branch was controlled by the Democrats. Interestingly, even during this period of divided government, there were no policy gaps and three policies were still enacted.

Block 5: 1972–1978 – This era was similar to the last, except the president was Gerald Ford rather than Richard Nixon until 1977. In 1977, Carter became president, and so there was no longer a divided government.

Block 6: 1978–1990 – During this era, there were numerous policy gaps. This was due to both Reagan's presidency, and his emphasis on shrinking the government through a lack of new policies, as well as divided government. During this period, there was only one enacted policy.

Block 7: 1990–2004 – Although Democratic President Bill Clinton led most of this era, divided government stifled housing policy. The beginning of the Republican Contract with America resulted in partisan politics during this period from both parties, stifling productivity with regard to housing markets. By 2001, Republicans controlled all branches of government, and in conjunction with the early 2000s recession, several new policies were enacted. President Bush's conservative ideology differed with past conservatives in that he was not completely opposed to expanding the government in certain areas, leading to several housing policies during this time.

Block 8: 2004–2018 – This period of time included the Great Recession of 2007–2009, the intense partisan politics of the early-to-mid 2010s, and the modern Trump era. At the end of Bush's presidency, a booming economy resulted in a lack of new policies. This all ended by 2008, when the Great Recession led both Bush and Obama to enact many new policies. By 2016, with the election of Trump, several new policies have been created, although the results of these policies are yet to be seen.

Finally, based on the data from Table 8.1, clearly, no matter the policy block, there have always been more supply-side policies than demand-side policies throughout U.S. history. Even in the 1930s, when there was an extreme number of housing policies enacted during the Great Depression, there were more supply-side policies than demand-side policies. With a high of five (5) supply-side policies, and only one (1) demand-side policy, even during a Democratic hold on government, there is clearly a supply-side policy bias in housing policy.

However, the supply-side policies have a range (high minus low number of policies) of four (4), with the demand-side policies having a range of three (3), and so even though there have been more supply-side policies than demand-side policies, the number enacted per decade have remained roughly constant at three (3), although the number of supply-side policies range exceeds demand-side policies.

Housing Policy Results by Political Party: President, Senate, and House

Table 8.2 shows a breakdown of supply-side and demand-side policies based on which political party was in power when the act was passed. The political parties are then divided by the Presidency, U.S. Senate, and U.S. House of Representatives. The results show that of the total housing policies enacted since 1930, 74 of the occurrences were of Democratic Party, 58 supply

Table 8.2 Housing Policies by Political Party: President, Senate and House

Housing Policy Breakdown by Democrat vs. Republican

		Pres.	Sen.	House	Total
All policies	Dem.	22	25	27	74
	Repub.	13	10	8	31
Ratio Dem-Rep		1.69×	2.50×	3.38×	2.39×
Supply-side policies	Dem.	17	20	21	58
	Repub.	11	8	7	26
Ratio Dem-Rep		1.55×	2.86×	3.00×	2.23×
Demand-side policies	Dem.	5	5	6	16
	Repub.	2	2	1	5
Ratio Dem-Rep		2.50×	2.50×	6.00×	3.20×

Source: Congressional Record

side and 16 demand side, significantly greater than the Republican Party.

Table 8.2 shows housing policies by political party for the president, senate, and house. The research shows that of all housing policies, supply-side and demand-side policies enacted; the Democratic Party – represented by the president, senate, and house – was more than two times (2×) more likely to be in power, as opposed to Republicans.

Supply Policy Descriptive Statistics

Since 1950, supply-side housing policy acceptance has slowed down significantly. From 1950 to 1960, 1971 to 1980, and 1991 to 2000, there were no supply-side policies. See Table 8.5d for breakdown by government branch party control during election periods since 1930 (Tables 8.3a and 8.3b).

Since 1930, and post-World War II, there have been multiple housing policies per decade. High supply-side policies occurred in the decades of 1930–1940 (5), 1941–1950 (4), 1961–1970 (7), and 2001–2010 (5), for an average of five policies per decade.

Table 8.3a Supply-Side Policies per Decade

Decade	Policies per Decade
Low (Years No Supply-Side Policies)	
1950–1960	0 policies
1971–1980	0 policies
1991–2000	0 policies
High (Years High Supply-Side Policies)	
1930–1940	5 policies
1941–1950	4 policies
1961–1970	7 policies
2001–2010	5 policies

Table 8.3b Policies per Decade

Stat	Policies per Decade
Mean	3.11
Median	3.00
Mode	4.00

Demand Policy Descriptive Statistics

Historically, demand-side housing policy has been absent in the United States. It wasn't until the early 2000s that demand-side policy started to increase in frequency (3). See Table 8.5c for breakdown on demand-side policy legislation by two-year election periods (Tables 8.4a and 8.4b).

Tables (8.5a–d) Election Periods (2 Years) by Party Control and Type of Housing Policy Passed/Not Passed

Of the total election periods (43) since 1930, when housing policy enacted, the Democrats controlled the Presidency 24 (56%), Senate 28 (65%), and House 31 (72%) of the time, whereas the

Table 8.4a Demand-Side Policies per Decade

Low (Years No Demand-Side Policies)

Decade	Polices per Decade
1940–1950	0 policies
1950–1960	0 policies
1960–1970	0 policies
1980–1990	0 policies
1990–2000	0 policies

High (Years High Demand-Side Policies)

2000–2010	3 policies

Table 8.4b Policies per Decade

Stat	*Policies Per Decade*
Mean	0.00
Median	0.77
Mode	3.00

Republican Party controlled the Presidency 19 (44%), Senate 15 (35%), and House 12 (28%) of the time.

Note: Election cycle periods broken down by two-year periods (total 43).

Of the total supply-side policies enacted during election periods since 1930, the Democrats enacted 13 (62%) under the Presidency, 15 (71%) Senate, and 17 (77%) through control of the House, whereas the Republican Party enacted 8 (32%) under the Presidency, Senate 6 (24%), and House 5 (29%).

Of the total demand-side policies enacted during election periods since 1930, the Democrats enacted 2 (50%) under the Presidency, 2 (50%) Senate, and 3 (75%) through control of the House, whereas the Republican Party enacted 2 (50%) under the Presidency, Senate 2 (50%), and House 1 (22%).

Table 8.5a Total Election Periods Where Housing Policy Enacted

		President	Senate	House
Control by government branch	Democrats	24 (56%)	28 (65%)	31 (72%)
	Republicans	19 (44%)	15 (35%)	12 (28%)
	Total election periods	43 (100%)	43 (100%)	43 (100%)

Table 8.5b Total Election Periods Where Supply-Side Housing Policy Enacted

		President	Senate	House
Supply side Total policies	Democrats	13 (62%)	16 (76%)	15 (71%)
	Republicans	8 (32%)	6 (24%)	5 (29%)
	Totals	21 (100%)	21 (100%)	21 (100%)

Table 8.5c Total Election Periods Where Demand-Side Housing Policy Enacted

		President	Senate	House
Demand side Total policies	Democrats	2 (50%)	2 (50%)	3 (75%)
	Republicans	2 (50%)	2 (50%)	1 (25%)
	Totals	4 (100%)	4 (100%)	4 (100%)

Table 8.5d Total Election Periods Where Producing No (Gap) Housing Policies Enacted

		President	Senate	House
Gap policy Total policies	Democrats	11 (50%)	13 (59%)	15 (68%)
	Republicans	11 (50%)	9 (41%)	7 (32%)
	Totals	22 (100%)	22 (100%)	22 (100%)

Of the total Gap Policy periods since 1930, the Democrats enacted 11 (50%) under the Presidency, 13 (59%) Senate, and 15 (68%) through control of the House, whereas the Republican Party enacted 11 (50%) under the Presidency, Senate 9 (41%), and House 7 (32%).

In summary, the Democratic Party controlled the Presidency, Senate, and House twice as often as the Republican Party. Likewise, the Democratic Party passed roughly two times as many supply-side policies. However, when it comes to demand-side policies, Democrats passed the same amount, roughly 50% when controlling all branches.

Table 8.6 Housing Policy and Political Party Combinations

Presidential/ Congressional Terms 1930–2018 (43)								
Policy Counts	*Democratic-Republican Combinations*							
Government branch combinations	PD SD HD	PD SR HD	PD SR HR	PD SD HR	PR SR HR	PR SD HR	PR SD HD	PR SR HD
Supply-side policies								
Yes	10	1	1	1	3	0	4	1
Total (21)	Total 13				Total 8			
No (Gap) Policy	6	0	4	1	2	0	6	3
Total (22)	Total 11				Total 11			
Demand-side policies								
Yes	1	1	0	0	1	0	1	0
Total (4)	Total 2				Total 2			

Table 8.6 (Legend)

Key	
PD	Democrat President
PR	Republican President
SD	Senate Democrat
SR	Senate Republican
HD	House Democrat
HR	House Republican

Below are the historical government branch combinations (organized by party President) between the two parties, broken down by two-year election periods. The table shows, with each respective combination, how many housing policies (supply and demand side) were enacted historically since 1930.

As shown in Table 8.6, since 1930, there have been 43 total two-year election periods. The results show, since 1930, there have been 21 supply-side housing policies enacted over the course of the 43 election periods, which is much larger than the total demand-side housing policies enacted (**4**).

In total, 13 supply-side housing policies have been passed under a Democratic Party President. Of the 13, ten **(77%)** of the supply-side policies have been enacted under control of a Democratic Party President, Senate, and House. Similarly, eight supply-side policies were enacted under a Republican Party President.

However, when it came to control of all three branches of the federal government for the Republican Party, there were only **three** policies enacted (38%) of the total eight. The majority of housing policy passed under a Republican Party President came from having a Democrat House and Senate (**four** policies).

When it came to no (gap policy) periods under party Presidents, Republicans and Democrats were the same (ten periods of no housing policy enacted). Under a Democratic Party President, **6 (60%)** of the blockage came from having control of all three branches of the federal government. For Republicans, only **2 (20%)** of the 10 housing policies blocked came from having control of all three branches.

It makes sense that Democrats have passed twice as many supply-side housing policies as Republicans because they have had more control of the federal government across all branches, nearly twice as many.

However, by organizing Table 8.6 by the Senate, Democrats have enacted **15** supply-side policies; while the total for a Republican Senate is **6** (see Table 8.5b). Similarly, counting the total policies enacted under each House, Democrats have enacted a total of **16**, while Republicans have enacted **5** (see Table 8.5c). The ratios for supply-side housing policies enacted per party change drastically when one counts policies by House and Senate.

Overall, Democrats have enacted more supply-side housing policy and show a bias when it comes to enacting supply-side housing policy. When comparing party Presidents with their opposing party-controlled House and Senate, a Republican President and Democratic Congress have enacted four supply-side housing policies while only blocking six (67% success ratio).

In contrast, under a Democratic Party President and Republican Congress, only one supply-side housing policy has been enacted while four have been blocked (25% success ratio). Historically, Democrats show a bias toward supply-side housing policy, and have had greater success at enacting supply-side policy under opposing party Presidents.

The prior section analyzed housing policy bias using political scientific policy analysis; the following sections show an econometric analysis of the various factors regarding outcomes from housing policy. The factors include affordability, CPI growth, and so on, and are correlated with each other to determine the relationship between the variables for forecasting purposes. The conclusion is supply-side housing policy bias which is reflected in a historical and forecast supply-demand gap.

9 Econometric Analysis and Results

Supply-Demand Correlation-Regression

After conducting the policy and political analysis proving supply-side bias, the next section looks at the econometrics of supply-side and demand-side factors, and the gap between the two.

Correlation Analysis

After reviewing the literature, an inventory of supply and demand data factors was compiled. After running the correlation analysis, and lagged time period correlation analysis, the following factors are considered significant in predicting future housing supply and demand.

Supply Factors

Factors showing strong correlation with housing supply are:

- Single- and multifamily housing starts (0 periods)
- CPI growth (−9 periods)
- Consumer confidence (−1 periods)
- Absolute change in GDP (0 periods)
- Change in home prices (−5 periods)
- Change in housing inventory (−4 periods)
- Interest Rates (0 periods)
- Affordability (0 periods)

DOI: 10.1201/9781003223436-9

Table 9.1 Supply Side Variable Correlation and Regression Results

Supply Variables	Correlation	
	Coef. (%)	*t-stats*
YOY Ann. Absolute Change GDP (0)	52.59	6.57
YOY Ann. Absol. Change Inventory (−4)	42.38	4.97
Rolling Ann. Qtr. Avg.10 year Int. Rts. (0)	28.56	−3.17
YOY Ann. % Change Personal Income per Household (−3)	47.98	5.81
Rolling Ann. Qtr. Avg. Housing Affordability Index (0)	28.17	3.07
YOY Ann. % Change CPI Ndx. (−9)	57.50	7.47
YOY Ann. % Change Home Price (−5)	49.91	6.12
YOY Ann. % Change Consumer Confidence Index (−1)	57.39	6.49
Rolling Ann. Qtr. Avg. Home Sales (0)	2.47	0.26
Rolling Ann. Qtr. Avg. Single-Family Starts (0)	79.40	13.88
Rolling Ann. Qtr. Avg. Multifamily Starts (0)	84.84	17.04

Note: see Table F in the Exhibits section for definitions of dependent and independent variables. Also, time periods are quarterly leads and lags. The parentheses indicate the number of periods (Table 9.1).

Note: numbers in parentheses are the number of lagged periods by quarter.

Note: see Table B in the Exhibits section for correlation results.

All time periods for historical graphs run from the fourth quarter of 1970 to the second quarter of 2001.

Factors showing strong correlation with housing demand are:

- Number of persons per household (0 periods)
- Home price appreciation (0 periods)
- Income per household growth (−2 periods)
- Inflation rate (0 periods)
- Employment growth (−1 periods)

Table 9.2 Demand Side Variable Correlation and Regression Results

Demand Variables	Correlation	
	Coef. (%)	t-stats
YOY Ann. Absolute Change GDP (0)	15.70	1.69
YOY Ann. Change EMPLOY (−1)	**21.50**	**2.33**
Rolling Ann. Qtr. Avg. UNEMP (−1)	**17.63**	**−1.91**
YOY Ann. Change POP24-65 (0)	10.10	1.07
Number of Persons per Hhld. (0)	**61.70**	**8.33**
YOY Ann. % Change Personal Income per Household (−2)	**51.90**	**6.45**
Rolling Ann. Qtr. Avg. Housing Affordability Index (−0)	**22.53**	**2.41**
YOY Ann. % Change Home Price (0)	**57.01**	**7.37**
Rolling Ann. Qtr. Avg.10 year Int. Rts.(0)	13.75	−1.47
YOY Ann. % Change CPI Ndx. (0)	**28.09**	**3.11**

The bolded variables are those that are statistically significant

Supply and Demand Model Results

To eliminate problems associated with multicollinearity, correlations were run on the independent variables (Table 9.2). Variables exhibiting high correlation with other significant variables, and did not having theoretical validity, were thrown out. The reduced form models are given below.

Note: see Tables C, D, and E in the Exhibits section for regression, correlation, t-stats, z-stats, and p-value results.

Theoretical Forecast Models

Housing (Supply) = a + (BS1)(XS1)+(BS2)(XS2)+...+(BSn)(XSn) + e
Housing (Demand) = a + (BD1)(XD1)+(BD2)(XD2)+...+(BDn)
                                    (XDn) + e
(Bn) = Beta or slope coefficient.
(Xn) = Actual independent supply or demand variables.

Reduced Form – Supply Model Forecast Equation

Of the significant supply variables, change in GDP (IVGDP), change in housing inventory (HOUSINGINV) lagged four

periods, mortgage interest rates (IVINTRT), and inflation
(IVCPI) lagged nine periods produced a statistically significant
forecast model and adheres to economic theory. This model is
used to forecast future housing supply (starts).

DVSTARTS = 975,792 + 1.44 (IVGDP(0)) + 0.11 (HOUSINGINV
(−4)) +

[8.16]   [4.06]

−14,269 (IVINTRT(0)) + 5,528,833 (IVCPI(−9))
[−1.53]   [8.86]

**Multiple R = 82%   R Square = 67%   Adj. R Square = 66%
F Stat = 55.7**

Reduced Form – Demand Model Forecast Equation

Of the significant demand variables, change in Gross Domestic
Product (GDP) (IVGDP), change in number of persons per household (IVPERSPERHHLD), and home sales price appreciation
(IVSAL$GTH) produced a statistically significant forecast model
and adheres to economic theory. This model is used to forecast
future housing demand (household formations).

DVHHLDS = −2,923,272 + 1.05 (IVGDP(0)) + 1,374,827
(IVPERSPERH-
HLD(0)) +

[4.70]   [6.86]

3,654,143 (IVSAL$GTH(0))
[−1.53]

**Multiple R = 75%   R Square = 56%   Adj. R Square = 55%
F Stat = 46.9**

*Note: Numbers in brackets are "t" statistics; and numbers in paren-
theses are numbers of lagged periods. Regression was run at a 90%
confidence level.*

Back-Testing Forecast Models

Based on model results, supply and demand projections for housing were completed, along with back-testing to see if predicted values matched up with actual values. From the back-test analysis, it has been determined that the supply model, and produced predicted values, does a good job of tracking actual supply. The forecast model can then be used to project future housing supply with confidence (Figure 9.1).

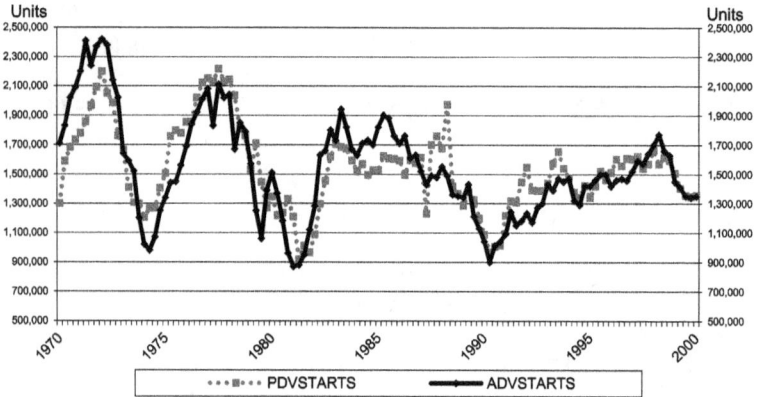

Source: Regional Financial Associates, Federal Reserve, U.S. Bureau of the Census.

Figure 9.1 Predicted vs. Actual Housing Supply.

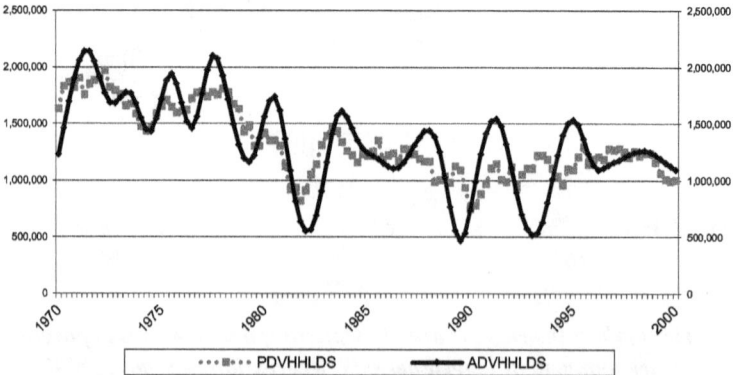

Source: Regional Financial Associates, Federal Reserve, U.S. Bureau of the Census.

Figure 9.2 Predicted vs. Actual Housing Demand.

From the analysis, it is also determined that the demand model, and produced predicted values, does a good job of tracking actual demand. The forecast model can then be used to project future housing demand with confidence (Figure 9.2).

Housing Supply and Demand Gap Analysis

After running the correlation analysis, determining predictive independent variables, eliminating variables that are highly correlated with each other, building the forecast models, and projecting housing supply and demand, the product of the two models produces the historical and projected housing gap for the United States. Based on the housing gap analysis, from 2000 through 2010, the gap between housing supply and demand will grow from 350,000 units per year to 480,000 units per year. This is an average annual compound growth rate over the ten-year period of roughly 1.0%.

The continuation of this housing-supply gap is a reflection of continued expansionary fiscal housing policy. Its long-term effect should be reflected in declining rates of real-home prices and higher housing affordability over the ten-year forecast period. The downside to this policy outcome is environmental degradation (urban sprawl) from predominantly suburban buildings.

However, since 2010 through 2018, it appears there has been a reduction in the positive effect of historical supply-side housing policy and its outcomes. This is reflected in rapidly rising housing costs (prices), and historically low housing production (relative to demand) and a collapse in housing affordability. This can be calculated through the equation: Total Housing Permits – Household Formations = Housing (–) Gap (National Low Income Housing Coalition, 2018).

The negative impact on housing production and affordability, due to the reduction in supply-side housing policy administration and legislation, has also exacerbated poverty levels and urban and rural homelessness (Figures 9.3 and 9.4).

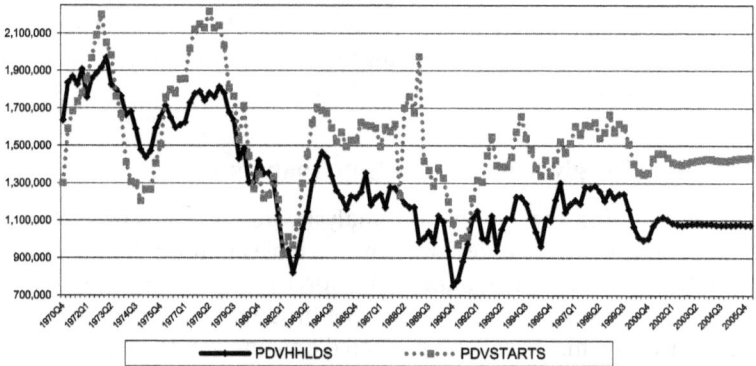

Source: Regional Financial Associates, Federal Reserve, U.S. Bureau of the Census.

Figure 9.3 Predicted Housing Demand vs. Supply.

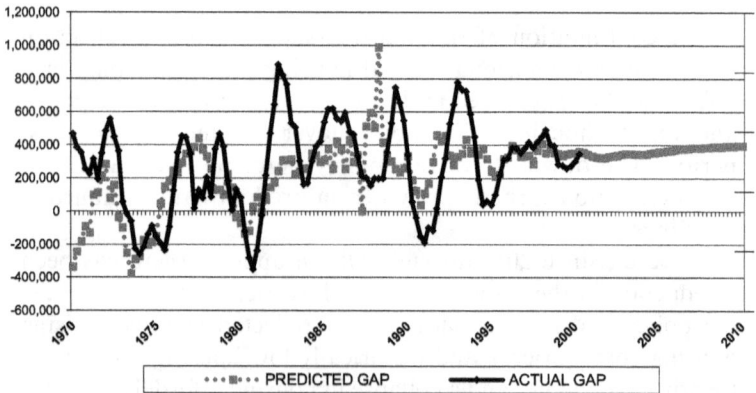

Source: Regional Financial Associates, Federal Reserve, U.S. Bureau of the Census.

Figure 9.4 Predicted vs. Actual Housing Gap.

10 Conclusion and Recommendations

Supply Side-Bias and Response to Housing Crisis

The federal government has acted to make mortgage funds more readily available, its direct role in housing finance has diminished markedly. In the past, federal policy actions have been of four different types:

(1) those which provide the framework and support for housing finance markets;
(2) those aimed at under-housed segments of the population;
(3) those which respond to crisis situations, for example, the creation of the HOLC in the 1930s;
(4) those that encourage housing production generally – such as the mortgage interest and real estate tax deductions allowed to homeowners.

Furthermore, some federal legislations aimed at non-housing problems, such as programs to improve urban transportation and environmental quality, have had substantial impact on the cost, location, and quality of residential structures.

Together with general economic growth, federal supply-side housing policies have been successful in nearly eradicating the worst housing problems of previous decades.

In the future, housing supply in the United States is expected to continue to outstrip housing demand. Housing supply is forecast to continue to exceed housing demand due to a federal housing policy bias toward supply-side legislation. However, over the past

DOI: 10.1201/9781003223436-10

ten years, this bias has deteriorated significantly, and has recently led to housing shortages.

Factors showing strong correlation with housing supply over the years are given as follows:

- Single and multifamily housing starts
- CPI growth
- Consumer confidence
- Absolute change in GDP
- Change in home prices
- Change in housing inventory
- Interest rates
- Affordability

From 2000 through 2010, the gap between housing supply and demand will grow from 350,000 units per year to 480,000 units per year. This is an average annual compound growth rate over the 28-year period of roughly 1.0%.

The continuation of this housing gap is a reflection of expansionary fiscal housing policy, and its effect should be reflected in declining rates of real home prices and higher housing affordability over the 28 years.

However, especially during the Obama administration, there was some transition toward demand-side legislation, as is often seen in recessions. This could merely be an anomaly, but nevertheless the supply-side legislation bias still remains throughout the 21st century.

Note: during the Trump Administration, there were no real housing policies initiated.

The fact that housing affordability has dropped so low over the last decade can be attributed to this split in supply-side and demand-side legislation, and how supply-side bias in legislation is still dominant. For demand-side legislation to be successful, or legislation in general, there cannot be a split in the legislation, as has been seen over the last decade.

There are still other factors that may determine housing affordability in the next few years. For example, the ongoing debates on immigration are crucial as more immigrants means

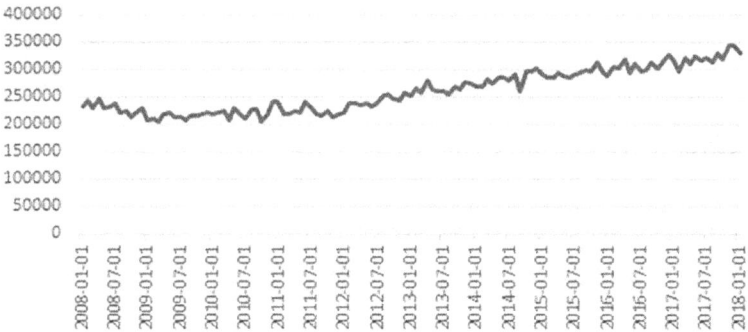

Figure 10.1 Median Price for Home Sales in the United States.
Source: Federal Reserve Bank of St. Louis.

more demand for housing, which could potentially drive prices up. However, fewer immigrants could mean less demand for housing and lower prices. Most importantly, the uncertainty about the number of immigrants creates some uncertainty about future housing demand, as cited by Miller, Mueller, and Dinn (2018) (Figure 10.1).

Percent Change in Median Home Prices:

2008–2010: −6.1%
2010–2012: 1.6%
2012–2014: 17.8%
2014–2016: 6.4%
2016–2018: 12.5%
2008–2018: 41.8%

House prices have been on an upward trend ever since 2008, and this can be attributed to cost push supply-side bias in legislation and also several demand-side policies (Figure 10.2).

Percent change in median price to buy a new home between 2000 and 2018: over 95%.

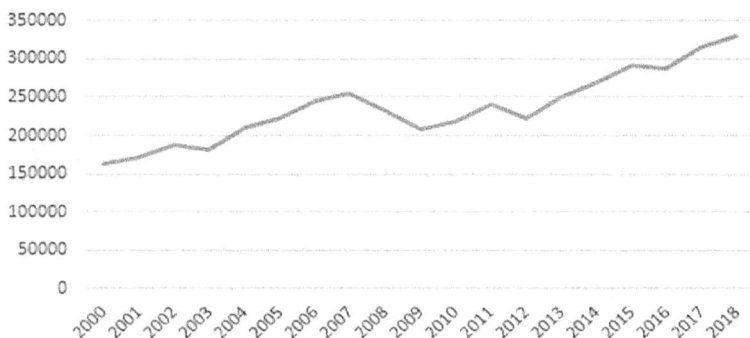

Figure 10.2 Median Prices to buy a New Home in the United States by Year.
Source: U.S. Census Bureau.

The period from Q1 2005 to Q2 2007 is the growth peak period for home prices; and affordability rose from 15% to 20%, which is an average of 17%. These were the lowest levels since 1996 to 2000. The period from Q3 2007 to Q1 2012 was the recession period.

Affordability rose from 15% to 43% during 2007–2001, high-to-moderate affordability, up 28 percentage points, or 2,800 basis points. Finally, during the Q2 2012–Q2 2018 recovery period, affordability fell from 43% to 25%, low-and-falling affordability, down 18 percentage points, or 18 basis points.

Although there continues to be a historical supply-side housing policy bias, the trend in affordability looks to continue on a downward secular trend line. Based on history, this affordability will reverse in the next recession from 2021 to 2023, not due to oversupply, but due to deficient aggregate demand.

Recommendations

Although there is a historical housing supply-side policy bias, in the last 20 years, these policy biases have declined and have been exposed to longer policy gaps.

Based on historical housing policy and economic outcomes, the following recommendations are made.

The downside to supply-side policies is environmental degradation and urban sprawl. To take environmental conditions into consideration, the federal government should look at including "smart growth" incentives. These incentives would channel housing development to existing infill or central city locations.

Other policy recommendations are to eliminate or reduce mortgage interest deductions, reduce subsidized mortgage finance operations, and scale back on government warranties. Due to current policies in place, the housing market in the United States could be exposed to oversupply conditions in the future. Although market oversupply is a risk to developers and property owners, it is a benefit to consumers due to lower prices and higher levels of affordability.

Historical federal housing policy is supply-side biased, and produces supply bubbles, because it benefits the general public. Overall, United States housing policy is effective in increasing homeownership, housing supply, construction quality, and better living conditions.

However, the supply-side policy biases have been compromised over the last 17 years as current and past administrations have tried to dismantle the housing capital markets and financial institutional support.

For example, a significant number of people are being barred from home ownership due to the recent scaleback in supply-side legislation. Essentially, the housing market previously consisted of supply shocks that would occur due to the supply-side legislation bias. When supply shocks would occur, there would be an oversupply of houses on the market, which would benefit homeowners, but not developers. Now, the opposite is true; without these supply shocks, developers are benefitting instrumentally while homeowners are being damaged.

In 2005, the homeownership rate, according to the Census Bureau, was at a high of 69.1% (U.S. Census Bureau, 2021). Now, in 2019, after the supply-side bias in legislation has been lessened, homeownership rate is at 64.2%. If we assume the current U.S. population to be approximately 326 million people, the difference in homeownership rate within the 13-year period has led to about 16 million people without homes in less than two decades.

Therefore, we strongly recommend the return of pure supply-side legislation bias to ensure as many people as possible are able to afford homes, as was possible in the second half of the 20th century.

Housing economists and policy analysts should monitor presidential and congressional political outcomes to be able to forecast housing policy, and its outcomes.

For example, historically over the past 75 years, housing policy innovation (blocks) was enacted under a Democratic Presidency, and Democratic-controlled Senate and/or House. During housing policy gap periods, or periods of little housing policy legislation, there was a Republican presidency and Republican-controlled Senate and/or House. Therefore, under the last Republican president, and Republican-controlled Senate and House, there was no real significant housing policy (supply side or demand side), other than short-term multiplier effects from tax cuts.

Based on this research, future housing policy innovation (blocks), supply side or demand side, is forecast not to occur until the government is controlled by the Democratic Party: President, Senate, and House. With the Senate and House, this could have occurred as early as 2018 or 2020, whereas the President could have occurred as early as 2020. Thus, the supply gap for housing could remain low through 2022, and recover to historical average from 2024 through 2028.

In 2021, a Democrat (Joe Biden) was elected as president, and a democratically controlled house and senate. During the 2021–2024 congressional and presidential election cycles, there is a high probability of significant supply-side housing policy enacted. For example, Biden-Harris housing policy initiatives include Section 8 housing choice vouchers, affordable housing construction, mitigation of discriminatory housing policies, reduction in exclusionary zoning policies, higher density housing, eviction moratorium and rental assistance, access to direct mortgage finance, first-time home buyer/renter tax credits, credit enhancements, focus on homelessness, etc.

Anticipated Usefulness of Results

Managerial or Academic Value of Research

The managerial value of the research is to know the history of housing policy in the United States by understanding how the policy evolved and its outcomes over time; this allows the manager to think strategically. If the developer or the owner of a housing unit, project, or subdivision understands how current federal housing policies affect housing market conditions in the future, the principal is in a better position to take advantage of market conditions, or programs and subsidies, offered by the government. The principal will also have the knowledge of what programs work and what programs do not work, and how not to be negatively impacted by federal policy mistakes.

This research also provides the manager the ability to forecast housing supply, demand, and gaps going forward. If the gap between housing demand and supply is severe, then the manager of a residential development company can plan for future increasing demand for his or her company's products and services. If the gap between housing demand and supply is less severe, or the market actually becomes oversupplied, the manager can plan for the decline in future demand, and reduce his or her development products and services accordingly.

Now investors can build and manage single-family and apartment portfolios, with the basic understanding of policy and economic fundamentals that drive demand and supply for housing. Since aggregate factors driving housing demand have been determined, related housing market demand and supply fundamentals can be analyzed, and factors determined for metro and submarket performance, and used in the selection process. Portfolio optimization and target allocations can also be determined, and hedging strategies established, using transaction-based housing indices for benchmarking and property derivative positions (single family).

The academic value from the research is to understand federal housing policy and its outcome on overall supply and its outcome on overall social well-being of the population. Looking at housing policy in a historical context provides academics the opportunity to see what programs and legislation were enacted at what periods of time, and under what political and economic conditions. This research provides academics the opportunity to analyze the history of housing policy in the United States and rationalize what policies have been effective in solving the housing crisis and what policies have been mistakes followed by their outcomes.

The academic value from the research is also to continue to test the significance of supply and demand variables identified in previous research, and to introduce new variables that may not have been available or thought of in prior research studies. The real academic value from this research is the final supply and demand forecast models developed. These models help guide other forecasters in building housing forecast models, and depending on the results of the gap analysis, these help policymakers in identifying prior supply-side or demand-side policies that have worked to bring the national housing market back into long-run equilibrium.

Housing economics and policy analysts should monitor presidential and congressional political outcomes to be able to forecast housing policy and its outcomes.

For example, over the past 75 years housing policy innovation (blocks) was enacted under a Democratic presidency, and Democratic-controlled Senate and/or House. During policy gap periods or periods of little housing policy legislation it was under a Republican Presidency, a Republican Senate, and/or Republican House. Therefore, under the prior Republican President, and Republican-controlled Senate and House, there was no real significant housing policy (supply or demand side), other than multiplier effects from tax cuts.

Summary

The best way to analyze housing in the United States is through legislation produced by congress. Congress, as well as presidents through executive orders, has the ability to influence markets through either supply-side or demand-side legislation.

As the name suggests, *supply-side legislation involves policies that promote the supply of housing, while demand-side legislation involves policies that promote demand for housing.* Overall, with regard to the history of U.S housing policy, *the research proves there has been supply-side legislation bias.* Demand-side policies do not occur often, generally only during recessions. We hypothesize that supply-side legislative bias has led to stable housing markets since the 1950s, allowing most Americans to become homeowners.

There has been clear supply-side policy bias, as over the 88 years of housing policy in the U.S. housing policy generally is created by Democrats, whether it be a Democratic president, Senate, and/or House. Nevertheless, there have been larger policy gaps in overall housing policy recently, and we believe this is the main reason for the current housing crisis.

In the last decade, *there has been a shift toward more demand-side legislation, or no legislation (policy gaps) at all,* even though the traditional institutional supply-side policy bias still remains. With fewer supply-side policies, the traditional supply shocks if the housing market has declined, leading to more benefits for housing developers and brokers, but only detriments (rising costs and falling inventory and affordability) for housing consumers. *The scaling back of supply-side legislation must be slowed, or reversed, to ensure the maximum benefits (affordability) for U.S. households.*

Citations

Federal Reserve Bank of St. Louis. (2021, April 23). *Median sales price of houses sold for the United States.* Received 2018, from https://fred.stlouisfed.org/series/MSPUS

Miller, N. G., Mueller, P., & Dinn, M. J. (2018, July 31). Housing demand and immigration trends. *Real Estate Issues, 42*(8).

U.S. Census Bureau. (2021, April 27). *Quarterly residential vacancies and homeownership, first quarter 2021.* Release Number: CB21–56. https://www.census.gov/housing/hvs/files/currenthvspress.pdf

Exhibits

History U.S. Housing Policy Legislation

Table A Legislation by Political Party and Congressional Composition

History of U.S. Housing Policy

Year	Act(s)		Supply	Demand	President	Senate	House	Recession
1930–1932	Federal Home Loan Bank System	The Public Works Administration	Yes	Yes	D	R	D	Yes
1932–1934	HOLC	Resettlement Administration, National Housing Act	Yes	No	D	D	D	No
1934–1936	National Housing Act		Yes	N/A	D	D	D	No
1936–1938	Housing Act		Yes	N/A	D	D	D	Yes
1938–1940	Policy Gap (WWII)		N/A	N/A	D	D	D	No
1942–1944	Servicemen's Readjustment Act		Yes	N/A	D	D	D	No
1944–1946	Farmers Home Administration		Yes	N/A	D	D	D	Yes
1946–1948	Federal National Mortgage Association		Yes	N/A	D	R	R	No
1948–1950	Housing Act		Yes	N/A	D	D	D	Yes
1950–1952	Policy Gap (Korean War)		N/A	N/A	D	D	D	No
1952–1954	Housing Act		Yes	N/A	R	R	R	Yes
1954–1956	Policy Gap (Divided Gov.)		N/A	N/A	R	D	D	No

(Continued)

History of U.S. Housing Policy

Year	Act(s)	Supply	Demand	President	Senate	House	Recession
1956–1958	Policy Gap (Divided Gov.)	N/A	N/A	R	D	D	Yes
1958–1960	Policy Gap (Divided Gov.)	N/A	N/A	R	D	D	No
1960–1962	Housing Act	Yes	N/A	D	D	D	Yes
1962–1964	Housing Act	Yes	N/A	D	D	D	No
1964–1966	Housing and Urban Development Act	Yes	N/A	D	D	D	No
1966–1968	Demonstration Cities and Metropolitan Development Act	Yes	N/A	D	D	D	No
1968–1970	Housing and Urban Development Act	Yes	N/A	R	D	D	Yes
1970–1972	Housing and Urban Development Act Emergency Home Finance Act	Yes	N/A	R	D	D	No
1972–1974	Policy Gap (Divided Gov.)	N/A	N/A	R	D	D	Yes
1974–1976	ECOA Housing and Community Development Act	Yes	Yes	R	D	D	Yes
1976–1978	Housing and Community Development Act	No	N/A	D	D	D	No-
1978–1980	Policy Gap (Focus on Economic Stagflation)	N/A	N/A	D	D	D	Yes
1980–1982	Policy Gap (Repub. Domination)	N/A	N/A	R	R	D	Yes

1982–1984	Policy Gap (Repub. Domination)		N/A	N/A	R	R	D	No
1984–1986	Policy Gap (Repub. Domination)		N/A	N/A	R	R	D	No
1986–1988	Tax Reform Act		Yes	N/A	R	R	D	No
1988–1990	Policy Gap (Divided Gov.)		N/A	N/A	R	D	D	No
1990–1992	National Affordable Housing Act		Yes	N/A	R	D	D	Yes
1992–1994	Policy Gap (Economic Boom)		N/A	N/A	D	D	D	No
1994–1996	Policy Gap (Divided Gov.)		N/A	N/A	D	R	R	No
1996–1998	Policy Gap (Divided Gov.)		N/A	N/A	D	R	R	No
1998–2000	Policy Gap (Divided Gov.)		N/A	N/A	D	R	R	No
2000–2002	Economic Growth and Reconciliation Act		Yes	N/A	R	R	R	Yes
2002–2004	American Dream Downpayment Act	JGTRRA	Yes	No	R	R	R	No
2004–2006	Policy Gap (War on Terror)		N/A	N/A	R	R	R	No
2006–2008	Policy Gap (War on Terror)		N/A	N/A	R	D	D	Yes
2008–2010	HERA, HASP, TRUIRJCA	TARP, ARRA	No/No/Yes	No/Yes	D	D	D	Yes
2010–2012	Policy Gap (Divided Gov.)		N/A	N/A	D	D	R	Yes
2012–2014	American Taxpayer Relief Act		Yes	N/A	D	D	R	No
2014–2016	Policy Gap (Divided Gov.)		N/A	N/A	D	R	R	No
2016–2018	TCJA	BBABP, MAHWA	No	No/Yes	R	R	R	No

Source: Congressional Record
R = Republican; D = Democrat

Table B Supply-Side Correlation Analysis

	IVSTARTS (0)	IVGDP (0)	HOUSINGINV (−4)	IVINTRT (0)	IVHHLDGTH (−3)	IVCPI (−9)	IVSal$Gth (−5)	IVSFSTARTS	IVMFSTARTS
IVSTARTS	100%								
IVGDP(0)	53%	100%							
HOUSINGINV (−4)	42%	14%	100%						
IVINTRT(0)	−29%	−29%	6%	100%					
IVHHLDGTH (−3)	48%	6%	42%	12%	100%				
IVCPI (−9)	58%	−2%	25%	−14%	74%	100%			
IVSal$Gth(−5)	50%	7%	26%	−22%	53%	78%	100%		
IVSFSTARTS	79%	69%	26%	−57%	31%	42%	42%	100%	
IVMFSTARTS	85%	21%	43%	6%	47%	52%	40%	35%	100%

Table C Master Supply Model Regression Summary Output

Regression Statistics

Multiple **R**	0.818198181
R Square	0.669448263
Adjusted **R** Square	0.6574282
Standard error	206070.2537
Observations	115

ANOVA

	df	SS	MS	F	Significance F
Regression	4	9.46021E+12	2.36505E+12	55.69423846	1.36737E-25
Residual	110	4.67114E+12	42464949446		
Total	114	1.41314E+13			

	Coefficients	Standard Error	t-Stat	P-value	Lower 95%	Upper 95%	Lower 90.0%	Upper 90.0%
Intercept	975792.3742	101924.6193	9.573667101	371566E-16	773801.607	1177783.141	806717.4879	1144867.26
IVGDP(0)	1 43674E-06	1.7616E-07	8.155899176	6 1531 E-13	1.08763E-06	1.78585E-06	1.14452E-06	1.72896E-06
HOUSINGINV(-4)	0.11434858	0.028144584	4.06289844	91196E-05	0.058572601	0.170124574	0.067661708	0.161035467
IVINTRT(0)	-14269.34515	9317.509464	-1.53145486	0.128527952	-32734.47064.	4195.780334	-29725.44266	1186.75237
IVCPI(-9)	5528833.711	624250.1483	8.856759947	1.61098E-14	4291715.855	6765951.566	4493313.308	6564354.114

Table D Demand-Side Correlation Analysis

	DVHHLDS	IVGDP	IVEMP (-1)	IVUNEMP (-1)	IVPOP (0)	IVPERSPERHHLD (0)	IVHHLDGTHH (-2)	IVAFFNDX (0)	IVSal$Gth (0)	IVINTRT (0)	IVCPI (0)
DVHHLDS	100%										
IVGDP	16%	100%									
IVEMP(-1)	21%	88%	100%								
IVUNEMP(-1)	-18%	-42%	-47%	100%							
IVPOP(0)	10%	18%	4%	-45%	100%						
IVPERSPERHHLD(0)	62%	-22%	-14%	3%	6%	100%					
IVHHLDGTHH(-2)	52%	12%	29%	11%	-27%	45%	100%				
IVAFFNDX(0)	10%	22%	16%	-53%	64%	20%	-44%	100%			
IVSal$Gth(0)	57%	-10%	11%	-7%	3%	50%	70%	-17%	100%		
IVINTRT(0)	-14%	-29%	-33%	67%	-67%	-13%	18%	-87%	-14%	100%	
IVCPI(0)	28%	-51%	-37%	19%	-19%	38%	59%	-57%	60%	37%	100%

Table E Master Demand Model
Regression Summary Output

Regression Statistics

Multiple R	0.747698053
R Square	0.559052379
Adjusted R Square	0.547134876
Standard Error	270401.8594
Observations	115

ANOVA

	df	SS	MS	F	Significance F
Regression	3	1.02898E + 13	3.42994E + 12	46.91019302	1.16876E-19
Residual	111	ε.11601E + 12	7311716560		
Total	114	1.84058E + 13			

	Coefficients	Standard Error	t Stat	P-value	Lower 95%	Upper 95%	Lower 90.0%	Upper 90.0%
Intercept	−2923272.785	550825.7109	−5.307073958	5.76348E-07	−4014770.357	−1831775.214	−3836926.171	−2009619.4
IVGDP	1.04474E-06	2.22371E-07	4.698185969	7.57337E-06	6.04097E-07	1.48538E-06	6.75894E-07	1.41359E-06
IVPERSPERHHLD(0)	1374827.863	200510.3341	6.856643422	4.17613E-10	977503.3768	1772152.35	104241.817	1707413.91
IVSal$Gth(0)	3654143.072	769889.3743	4.746322256	6.21903E-06	2128556.365	5179729.779	2377129.273	4931156.871

Table F Housing Supply and Demand Analysis Correlation and Regression Statistical Results

Summary Analysis of Significant Independent Variables

	Correlation Coef.	t-Stats	F Stats	P-value
Demand Variables				
Independent variables were regressed against housing demand.				
Household formations were used as a proxy for total housing demand.				
YOY Ann. Absolute Change GDP (0)	15.70%	1.69	2.87	0.093
YOY Ann. Change EMPLOY (–1)	21.50%	2.33	5.46	0.021
Rolling Ann. Qtr. Avg. UNEMP (–1)	17.63%	–1.91	3.66	0.058
YOY Ann. Change POP24-65 (0)	10.10%	1.07	1.17	0.282
Number of persons Per. Hhld. (0)	61.70%	8.33	69.47	0.000
YOY Ann% Change Personal Income Per Household (–2)	51.90%	6.45	41.42	0.000
Rolling Ann. Qtr. Avg. Housing Affordability Index (0)	22.53%	2.41	5.83	0.017
YOY Ann% Change Home Price (0)	57.01%	7.37	54.42	0.000
Rolling Ann. Qtr. Avg. 10 year Int. Rts. (0)	13.75%	–1.47	2.17	0.143
YOY Ann% Change CPI Ndx. (0)	28.09%	3.11	9.67	0.002
Supply Variables				
Independent variables were regressed against housing supply.				
Housing starts were used as a proxy for total housing supply produced.				
YOY Ann. Absolute Change GDP (0)	52.59%	6.57	43.21	0.000
YOY Ann. Absol. Change Housing Inventory (–4)	42.38%	4.97	24.74	0.000
Rolling Ann. Qtr. Avg. 10 year Int. Rts. (0)	28.56%	–3.17	10.04	0.002
YOY Ann% Change Personal Income Per Household (–3)	47.98%	5.81	33.78	0.000

Rolling Ann. Qtr. Avg. Housing Affordability Index (0)	28.17%	3.07	9.39	0.003
YOY Ann% Change CPI Ndx. (−9)	57.50%	7.47	55.82	0.000
YOY Ann% Change Home Price (−5)	49.91%	6.12	37.49	0.000
YOY Ann% Change Consumer Confidence Index (−1)	57.39%	6.49	42.23	0.000
Rolling Ann. Qtr. Avg. Home Sales (0)	2.47%	0.26	0.68	0.793
Rolling Ann. Qtr. Avg. Single-Family Starts (0)	79.40%	13.88	192.78	0.000
Rolling Ann. Qtr. Avg. Multifamily Starts (0)	84.84%	17.04	290.25	0.000

Notes: "t" and "F" Stats were produced at a 90% confidence level.
YOY is year over year on a quarterly basis.
(X) is the number of lagged periods, (0) being not lag, and (−1) being one period lag.

References

Abraham, J. M., & Hendershott, P. H. (1996). Bubbles in metropolitan housing markets. *Journal of Housing Research, 7*(2), 191–207.

American Legislative Exchange Council. (1995, January 4). *The homeownership and opportunity for people everywhere act.* The Heartland Institute. https://www.heartland.org/publications-resources/publications/the-homeownership-and-opportunity-for-people-everywhere-act

American Planning Association. (n.d.). *Housing act of 1949.* Retrieved 2018, from https://www.planning.org/awards/2014/1949housingact.htm

ATTOM Staff. (2018, May 21). *Americans departing high-priced housing markets for more affordable alternatives.* ATTOM Data Solutions. https://www.attomdata.com/news/market-trends/home-sales-prices/home-affordability-top-departure-and-arrival-markets/

Betters, P. V. (1949). The United States housing act 1949. *Annals of Public and Cooperative Economics, 20*(3), 375–378. https://doi.org/10.1111/j.1467-8292.1949.tb01586.x

Blount, J. (2016). *Housing, not unemployment, major concern when World War II veterans returned home.* Historical Collection at the Lane. Hamilton Journal-News. https://sites.google.com/a/lanepl.org/columns-by-jim-blount/2016-articles/housing-not-unemployment-major-concern-when-world-war-ii-veterans-returned-home

Bratt, R. G., & Keating, W. D. (1993). Federal housing policy and Hud: Past problems and future prospects of a beleaguered bureaucracy. *Urban Affairs Quarterly, 29*(1), 3–27. https://doi.org/10.1177/004208169302900101

Carn, N., Rabianski, J., Racster, R., & Seldin, M. (1988). *Real estate market analysis: Techniques and applications.* Prentice Hall.

Cassidy, J. (2017, January 9). Obama's economic record: An assessment. *The New Yorker.* https://www.newyorker.com/news/john-cassidy/obamas-economic-record-an-assessment

CNN Editorial Research. (2017, August 27). *American generation fast facts.* CNN. Retrieved 2017, August 27 from https://www.cnn.com/2013/11/06/us/baby-boomer-generation-fast-facts/index.html

Couch, L. M. (2020, Summer). *Falling short: Federal housing assistance is failing older adults.* Generations American Society on Aging. https://generations.asaging.org/federal-housing-assistance-older-adults-failing

Federal Reserve Bank of St. Louis. (2021, April 23) *Median sales price of houses sold for the United States.* Received 2018, from https://fred.stlouisfed.org/series/MSPUS

Finegold, K., Schardin, S., & Wherry, L. (2004, April 21). *Block grants: Historical overview and lessons learned.* The Urban Institute. New Federalism: Issues and Options for States, Series A, (A-63). http://webarchive.urban.org/UploadedPDF/310991_A-63.pdf

Fish, G. S. (1979). *The story of housing.* Macmillan Publishing.

Ginnie Mae. (n.d.). *Our history.* Retrieved 2018, from https://www.ginniemae.gov/about_us/who_we_are/Pages/our_history.aspx

Gissler, S., & Narajabad, B. (2017, October 28). *The increased role of the federal home loan bank system in funding markets, part 1: Background.* Board of Governors of the Federal Reserve System. https://www.federalreserve.gov/econres/notes/feds-notes/the-increased-role-of-the-federal-home-loan-bank-system-in-funding-markets-part-1-background-20171018.htm

Gross, T. (2017, May 3). *A 'forgotten history' of how the U.S. government segregated America.* NPR. https://www.npr.org/2017/05/03/526655831/a-forgotten-history-of-how-the-u-s-government-segregated-america

Hanushek, E. A., & Quigley, J. M. (1980, February 22). What is the price elasticity of house demand? *Review of Economics and Statics, 62*(3), 449–454.

Harrington, D. E. (1989). An intertemporal model of housing demand: Implications for the price elasticity. *Journal of Urban Economics, 25*(2), 230–246.

Haurin, D. R., & Lee, K. (1989, November). A structural model of the demand for owner-occupied housing. *Journal of Urban Economics, 26*(3), 34–360.

HUD Exchange. (1999, January). *Housing and community development act of 1974.* https://www.hudexchange.info/resource/2184/housing-and-community-development-hcd-act-of-1974/

IRS. (n.d.). *Credits and deductions for individuals.* Retrieved 2018, from https://www.irs.gov/credits-deductions-for-individuals

Jacobs, B. G. (1986). *Guide to federal housing programs.* Bureau of National Affairs.

Jan, T., Dewey, C., & Stein, J. (2018, April 25). HUD Secretary Ben Carson to propose raising rent for low-income Americans receiving federal housing subsidies. *The Washington Post.*

John F. Kennedy Presidential Library and Museum. (2018). *Legislative summary: Housing.* Retrieved 2018, from https://www.jfklibrary.org/ Research/Research-Aids/Ready-Reference/Legislative-Summary-Main-Page/Housing.aspx

Johnson, B. (1982). *Resolving the housing crisis: Government policy, decontrol, and the public interest.* Pacific Institute for Public Policy Research.

Keith, N. S. (1973). *Politics and the housing crisis since 1930.* Universe Books.

Lerner, M. (2018, October 4). 10 years later: How the housing market has changed since the crash. *The Washington Post.* https://www.washingtonpost.com/news/business/wp/2018/10/04/feature/10-years-later-how-the-housing-market-has-changed-since-the-crash/

Marcin, T. C., & Kokus, J. (1975). Economic and demographic factors affecting demand into the 1980s and 1990s. *Real Estate Economics, 3*(3), 81–93.

Markels, A. (2008, February 28). *Comparing today's housing crisis with the 1930s.* U.S. News & World Report. https://money.usnews.com/money/personal-finance/real-estate/articles/2008/02/28/comparing-todays-housing-crisis-with-the-1930s

Massey, D. S. (2015). The legacy of the 1968 fair housing act. *Sociological Forum, 30*(S1), 571–588. https://doi.org/10.1111/socf.12178

McManus, J. O. (2015, December 25). *5 ways to protect your estate from capital gain taxes.* MarketWatch. https://www.marketwatch.com/story/5-ways-to-protect-your-estate-from-capital-gains-taxes-2015-12-25

Michel, N., & Ligon, J. (2015, May 11). *The federal housing administration: What record of success?.* The Heritage Foundation. https://www.heritage.org/housing/report/the-federal-housing-administration-what-record-success

Migration Policy Institute. (2019). *Legal immigration to the United States, 1820-present.* https://www.migrationpolicy.org/programs/data-hub/charts/Annual-Number-of-US-Legal-Permanent-Residents

Mitchell, J. P. (Ed.). (1985). *Federal housing policy and programs: Past and present.* Routledge.

National Archives. (2018, April 25). *The Civil Rights Act of 1964 and the equal employment opportunity commission.* https://www.archives.gov/education/lessons/civil-rights-act

82   *References*

Nigro, C. (2018, June 7). *Tenement homes: The outsized legacy of New York's notoriously cramped apartments.* New York Public Library. https://www.nypl.org/blog/2018/06/07/tenement-homes-new-york-history-cramped-apartments

Pozdena, R. J. (1988). *The modern economics of housing: A guide to theory and policy for finance and real estate professionals.* Quorum Books.

Rohn, A. (2014, January 22). *How much did the Vietnam War cost?* The Vietnam War. https://thevietnamwar.info/how-much-vietnam-war-cost/

Smith, W. F. (1970). *Housing: The social and economic elements.* University of California Press.

Social Welfare History Project. (2018). Tenement house reform. *Social Welfare History Project.* http://socialwelfare.library.vcu.edu/issues/poverty/tenement-house-reform/

Statista Research Department. (2017). *Number of housing units in the United States from 1975 to 2017 (in millions).* Statista. Retrieved 2017, from https://www.statista.com/statistics/240267/number-of-housing-units-in-the-united-states/

Stewart, E. (2017, July 19). *Fannie Mae and Freddie Mac would be privatized under proposed House budget.* https://www.thestreet.com/story/14233125/1/fannie-mae-and-freddie-mac-would-be-privatized-under-proposed-house-budget.html

The International Garden Cities Institute. (n.d.). *A prospectus for the international garden cities institute.* https://www.gardencitiesinstitute.com/sites/default/files/documents/igci_prospectus_low_res.pdf

The Library of Congress. (n.d.). *Congressional Record.* Congress.Gov. https://www.congress.gov/congressional-record

United States Census Bureau. (2018, May 24). *Census Bureau reveals fastest-growing large cities.* https://www.census.gov/newsroom/press-releases/2018/estimates-cities.html

United States Census Bureau. (n.d.). *Subject definitions.* https://www.census.gov/programs-surveys/cps/technical-documentation/subject-definitions.html#householdfamily

U.S.A. (2018). Federal home loan mortgage association (Freddie Mac). http://www.freddiemac.com/

U.S. Department of Housing and Urban Development (HUD). (2014, December 23). *Enforcement of Davis-Bacon Act wage requirements for projects participating in the second component of the Rental Assistance Demonstration program.* https://www.hud.gov/sites/documents/NOTDAVISBACENFSTRAT1223.PDF

U.S. Department of Housing and Urban Development (HUD). (n.d.-a). *History of fair housing.* Retrieved 2018, from https://www.hud.gov/ program_offices/fair_housing_equal_opp/aboutfheo/history

U.S. Department of Housing and Urban Development (HUD). (n.d.-c). *HOPE VI.* Retrieved 2018, from https://www.hud.gov/ program_offices/public_indian_housing/programs/ph/hope6

U.S. Department of Housing and Urban Development. (n.d.-d). *The Federal Housing Administration (FHA).* Retrieved 2018, from https:// www.hud.gov/program_offices/housing/fhahistory

U.S. Department of Veterans Affairs. (n.d.). Education and Training, Retrieved from 2018, https://benefits.va.gov/gibill/

U.S. HUD. (2020). *Picture of subsidized households.* https://www.huduser. gov/portal/datasets/assthsg.html

Van Vliet, W. (Ed.). (1998). *The encyclopedia of housing.* Sage Publications, Inc.

Weber, W., & Devaney, M. (1996). Can consumer sentiment surveys forecast housing starts? *Appraisal Journal, 4,* 343–350.

Western Regional Advocacy Office. (n.d.). History of slashing HUD budget sources. http://weap.org/uploads/fact%20sheets/SourcesWRAP-HistoryofSlashingHUDBudgetFactSheet-Final.pdf

Worley, J. (2019, October 18). *Home budget: Cost-of-living reality check.* Quicken. https://www.quicken.com/home-budget-cost-living-reality-check

Index

For Product Safety Concerns and Information please contact our EU
representative GPSR@taylorandfrancis.com
Taylor & Francis Verlag GmbH, Kaufingerstraße 24, 80331 München, Germany

www.ingramcontent.com/pod-product-compliance
Lightning Source LLC
Chambersburg PA
CBHW071057280326
41928CB00050B/2535